ADVANCE PRAISE FOR

RHETORIC ONLINE

"Barbara Warnick is the foremost rhetorical critic who has taken seriously the need to adapt rhetorical theory to the study of public communication on the Internet. In this book she advances her previous work by developing a critical approach specific to the online environment. The book is smart, focused, and well informed. Its attention to research in multiple fields—public communication, composition studies, technical communication, human-computer interaction, literary criticism—will make the book useful to scholars and students in these and other fields."

Carolyn R. Miller, SAS Institute Distinguished Professor
of Rhetoric and Technical Communication,
North Carolina State University

"Barbara Warnick has once again focused her sharp analytical lens on the rhetoric of digital discourse. With its emphasis on the public sphere, credibility, and interactivity, this new book provides us with a sound framework for both a critical and a heuristic view of the Internet. Using the concepts and examples presented here, we understand the Internet in all of its persuasive subtlety and we see ways in which our design choices—textual, visual, political—could be more adaptive to this new rhetorical space. This is an important book for rhetorical studies and Internet studies alike."

Laura J. Gurak, Professor and Head, Department of Writing Studies,
University of Minnesota; Author of
Persuasion and Privacy in Cyberspace *and* Cyberliteracy

RHETORIC
ONLINE

POLITICAL COMMUNICATION

FRONTIERS IN

Lynda Lee Kaid and Bruce Gronbeck
General Editors

Vol. 12

PETER LANG
New York • Washington, D.C./Baltimore • Bern
Frankfurt am Main • Berlin • Brussels • Vienna • Oxford

Barbara Warnick

RHETORIC ONLINE

Persuasion and Politics on the World Wide Web

PETER LANG
New York • Washington, D.C./Baltimore • Bern
Frankfurt am Main • Berlin • Brussels • Vienna • Oxford

Library of Congress Cataloging-in-Publication Data

Warnick, Barbara.
Rhetoric online: persuasion and politics on the world wide web / Barbara Warnick.
p. cm. — (Frontiers in political communication; v. 12)
Includes bibliographical references and index.
1. Rhetoric—Political aspects—United States. 2. Internet—United States.
3. Persuasion (Rhetoric). 4. Intertextuality. I. Title.
P301.5.P67W37   808'.0420285—dc22   2006026621
ISBN 978-0-8204-8802-8
ISSN 1525-9730

Bibliographic information published by **Die Deutsche Bibliothek**.
Die Deutsche Bibliothek lists this publication in the "Deutsche
Nationalbibliografie"; detailed bibliographic data is available
on the Internet at http://dnb.ddb.de/.

Cover design by Lisa Barfield

The paper in this book meets the guidelines for permanence and durability
of the Committee on Production Guidelines for Book Longevity
of the Council of Library Resources.

© 2007 Peter Lang Publishing, Inc., New York
29 Broadway, 18th floor, New York, NY 10006
www.peterlang.com

Printed in the United States of America

Contents

Preface

The present project was inspired by the work of others interested in how new communication technologies affect the form of the messages they convey. Among these works are Lev Manovich's *Language of New Media*, N. Katherine Hayles's *Writing Machines*, and Paul Messaris and Lee Humphreys's recent edited collection, *Digital Media: Transformations in Human Communication*. Manovich's groundbreaking study identified the unique characteristics of the digital media, noting the extent to which new technology has been used to preserve the appearance of conventional analog media forms that preceded it. Hayles's work examined the evolving forms of literature in an age when literary works are presented in multimedia form; narratives appear in hypertext; and artists' books combine the verbal and the visual in unprecedented ways. Essays in the Messaris and Humphreys's collection considered the nature and influence of computerized forms of communication in video, music, gaming, and virtual experience.

I agree with these authors that it is important to describe and analyze how the development of digital and other media technologies have changed the ways in which we communicate. I wrote this book because I am concerned about the need to preserve a record of digital rhetoric's first decade and also about the ongoing need to catalog and describe politics and persuasion in online political discourse, alternative media, new social movement activism, and other sites of online public persuasion. Like Manovich, I feel that it is important to document persuasive forms in new media environments, just as they are "coming into being . . . when the elements of previous cultural forms shaping [them are] still clearly visible and recognizable" (7). For this reason, this book takes a media ecology approach to explain how digitization has reshaped persuasion in public discourse during the past decade.

This book proposes a program for the continued study of online persuasive communication by rhetorical critics and analysts. It begins with an argument for the need to understand how new media forms have reshaped public persuasion, and it argues for greater attention to persuasive elements of online resistive discourse alongside the study of institutionalized discourse. It then moves on in Chapter 2 to describe how digitization, media convergence, and technological and social factors have changed the communication environment in which persuasion occurs. The remaining chapters then each explore online credibility, interactivity, and intertextuality to develop theories of persuasive action adapted to online discourse. These theories take into account such factors as coproduction of messages, uncertain authorship, discontinuous modular texts, hypertext structure, and multimediated communication as they are found online. In each of the three chapters, I have included a case study to clarify how the theory might be applied. My hope is that this book might provide some footing to enable other critics and theorists to develop further the models and approaches well suited to the study of online persuasion.

This work is the outcome of a decade of research on the forms of online persuasion, and it has been deeply influenced by the ideas and work of my colleagues and research assistants. The work by Kirsten Foot, my colleague at the University of Washington, has shaped my thinking on the use of discourse in political campaigning. Suggestions from Crispin Thurlow contributed to the theoretical development of Chapter 3 on credibility. Without the support of my former department at Washington and its chair Jerry Baldasty, the work on the book could not have proceeded as well as it has. In particular, I want to acknowledge the contributions of my former research assistants. Danielle Endres and Dru Williams contributed measurably to the study of rhetorical dimensions of interactivity and to the user effects study cited in Chapter 4. Sara J. Morgan and Jonathan Moore developed materials supporting Chapter 3's theory on coproduction and online interactivity. Timothy Pasch's assistance in identifying intertextual allusions and in archiving examples in Chapter 5 was invaluable. Finally, my editor at Peter Lang Publishing, Damon Zucca, has been consistently helpful in facilitating and overseeing the production of this book. The careful reading and numerous helpful comments and suggestions of the series editor Bruce E. Gronbeck have furthered the development of ideas and case studies in the book as a whole.

I also want to thank my husband, Michael R. O'Connell, for his long-standing support of my work, both intellectually and as "technical advisor."

Barbara Warnick
University of Pittsburg

Acknowledgments

Portions of Chapter 2 were adapted from my article "Looking to the Future: Electronic Texts and the Deepening Interface." *Technical Communication Quarterly* 14.3 (2005): 327–333, by permission from Lawrence Erlbaum Associates.

A portion of Chapter 4 was taken from my article, coauthored with Danielle Endres, "Text-Based Interactivity in Candidate Campaign Web Sites: A Case Study from the 2002 Elections." *Western Journal of Communication* 68.3 (2004): 322–342.

Figure 5.1, "GatesWay," is used by permission of www.organique.com

Figure 5.2, "The Benetton," and 5.3, "The Blackspot Sneaker," are used courtesy of www.adbusters.org

List of Illustrations

1

The Internet and the Public Sphere

During the past 20 years, we have witnessed increasing concern among sociologists, political scientists, and communication scholars about the health of the public sphere as a space for democratic deliberation and debate. To some extent, the present angst about the public sphere has its origin in Jürgen Habermas's 1962 book, *The Structural Transformation of the Public Sphere* (translated and published in English in 1989). In this work, Habermas presented a genealogy of the bourgeois public sphere, which had its origins in the seventeenth and eighteenth centuries. Using extensive citations from published documents of the time, he traced the rise of the bourgeois public sphere as an artifact of a combination of factors, including increasing literacy, a rising commercial and trade sector, and formation of a bourgeois class.

Habermas noted the importance of the private sphere—the realm of commodity exchange and social labor imbedded in the conjugal family—as it related to salons, the theater, reading practices, and social gatherings. It served as a resource for the public sphere, a space in which people read, discussed, and wrote about opinions, issues, and ideas in coffee houses and public meeting places. He described the characteristics of this bourgeois public sphere, including widespread rational-critical debate, equality and association among persons of unequal status, freedom from censorship of free expression, and the opportunity to reach consensus on what was practically necessary, as in the interest of all people.

The public sphere as described by Habermas was an idealized construct against which were measured the venues and practices for public discourse in other eras. It excluded many groups of people, especially the illiterate, women, laborers, and other groups in society (Fraser). Its exclusive emphasis on debate and rational-critical discussion also elided other forms of communication, such as performative and aesthetic expression, which unified social groups and enabled identity formation (Warner).

Nevertheless, the bourgeois public sphere served Habermas well in his purpose, which was to describe its decline from the nineteenth to the mid-twentieth century. He noted that during this time, the newspaper developed into a capitalist undertaking that was dependent on advertising revenues and thus became "enmeshed in a web of interests extraneous to [its] business that sought to exercise influence upon it" (185). To increase circulation and appeal to the mass public, newspapers lowered costs and reading levels and voiced political topics and issues. The late nineteenth and early twentieth centuries witnessed technological developments, such as the telegraph, telephone, and radio, which made possible the organizational and economic interlocking of the press and the homogenization of news sources. Media content itself was designed to stimulate consumption and was channeled into patterns that had mass appeal; specifically, it exploited events to attract audience attention. In addition, the twentieth century also brought the rise of public relations "where the addressee is 'public opinion' and the sender of the message hides his business intentions in the role of someone interested in the public welfare" (193). Habermas implied that the impacts of mass media, widespread capitalist influence, and the public's declining interest in public deliberation led to a disintegration of the public sphere. He concluded that

> [t]he old basis for a convergence of opinions has . . . collapsed. A new one is not brought about merely because the private interests inundating the public sphere hold on to its faked version. For the criteria of rationality are completely lacking in a consensus created by sophisticated opinion-molding services under the aegis of a sham public interest. (195)

Thus, in his conception of the public sphere, Habermas emphasized the importance of open public discourse to apply the will of the people to formulation of the policies and laws that govern them.

The mere existence of public communication is not enough to ensure the viability of a public sphere, however. Its continuance relies as well on the extension of a common culture, shared experience, communal values, skill in and commitment to the processes of deliberation, access to news and information, and the public's ability to influence social institutions and government policy.[1] A number of observers have recently argued that these dimensions of the public sphere have continued to decline during the past 50 years and that the idealized historical public sphere envisioned by Habermas cannot be sustained in the twenty-first century

(Bruner; Calhoun; Castells's *The Power*; Kellner). In various publications, these authors describe at least four current developments leading to the continuing disintegration of the public sphere: loss of nation-state sovereignty, corporatization of media, constraints on public political discourse, and the rise of scandal politics. At the same time, these critics express the hope that mobilization, resistance, and social activism as they commonly occur in community media (local radio and government-sponsored public television), alternative media, and computer-mediated communication might increase public awareness of major social and political issues.

The major purpose of the present chapter is to argue that a good deal of vibrant and effective public discourse in the forms of social activism and resistance occur online, that such discourse has had noticeable effects on society, and that it is therefore worthy of careful study by rhetoricians. Before turning to this more hopeful scenario, however, I will begin by describing what critics view as a major crisis in the contemporary public sphere.

Since the 1980s, a number of trends have contributed to the decline of the nation-state as a centered source of power and governance (Bruner; Castells's *The Power*; Fürsich and Robins). The growth of transnational corporations has suppressed regulations that formerly supported the wage and welfare needs of citizens in democratic capitalist states. Global corporations at present "can choose their sites of investment, production, tax, and residence as they wish" (Brown 348). In efforts to attract investment and promote economic growth, governments have abandoned state functions or found ways to perform them more cheaply or efficiently by contracting them out or leaving it to users to pay for them. Recent U.S. examples of this phenomenon include the defunding of higher education by state governments and President George W. Bush's failed proposal to privatize social security. Historical examples include the deregulation of the airline industry and telecommunications. As governments vacate their fiduciary commitments to protect the public interest, market forces move in to convert many functions to for-profit enterprises. Thus, as Richard Harvey Brown has noted, "the balance of class forces is shifting away from individual nation states and their respective forces toward an increasingly mobile and interdependent global capitalist class" (351). With regard to international trade practices as well as the economy and income distribution, policies are developed by entities such as the World Bank, the International Monetary Fund, and the World Trade Organization rather than by nation-states.[2]

A second development harmful to the public sphere is the growth of big media corporations through buyouts and consolidations. In the past two decades, the number of major corporations controlling interests in television, movies, music, radio, cable, and publishing has decreased from fifty to less than two dozen, with much of the control concentrated in less than ten (McChesney). For example, TimeWarner's holdings include Warner Brothers, AOL, CNN, HBO Turner, New Line Cinema, Castle Rock Entertainment, and others. Viacom's holdings include CBS and UPN Networks, MTV, Showtime, Paramount Pictures, Blockbuster Video, over 175 radio stations, and over 35 television stations. The same pattern of extensive media control holds for General Electric and Disney ("Who Owns the Media?"). These consolidations were made possible by the federal government's relaxation of media ownership rules.

The growth in large corporate media interests is harmful to the public sphere because, as Manual Castells has noted, "political communication and information are essentially captured in the space of media" (370). Media "captures" politics by becoming the primary means through which citizens experience political discourse. Since media coverage is profit driven, media content focuses on encapsulated news that holds viewer interest rather than on in-depth consideration of complex issues. During campaign seasons, media discourse on politics emphasizes incessant reportage of poll data, treating campaigns as "horse races" (Bimber and Davis; Jamieson). The public often gets political news through leaks and counter leaks to the press and disclosures of improprieties by candidates and elected officials. The media also focus on the visual campaign story, be it Bill and Hillary Clinton's response to Gennifer Flowers's allegations, Michael Dukakis in a tank, or a screaming Howard Dean. Castells has maintained that the media's reliance on visuals, sound bites, and encapsulated news has led to "an extreme simplification of political messages and polarized discussion of issues" (381).

The situation is made worse by the ubiquitous presence of political advertising. As Brown noted, "political disaffection is . . . fostered by the financing of elections by corporations and wealthy individuals who, in the United States for example, pay for about 80 percent of both Republican and Democratic campaigns" (350). Much of these funds is used to support mediated political ads that look like news in the form of "advertorials" and "infomercials" (Hamelink).

The third aspect of the current political climate harmful to the public sphere consists of constraints on political discourse. Public discourse on politics is limited not only by media practices but also by the behavior of political campaigns themselves. Due to the possibility of negative media coverage or attacks by their candidates' opponents, campaigns seek to carefully control campaign discourse. In a study of the lack of interactivity on political campaign Web sites, Jennifer Stromer-Galley found that campaigns favor the crafting of broad, ambiguous policy statements that can be interpreted by voters in ways that bring the message as close as possible to a voter's own position. Having developed such broad policy statements, candidates then work to "stay on message" and to focus voter attention in their campaign within a given framework. Furthermore, campaigns design political ads to focus primarily on negative messages so as to undermine their opponent's message while advancing their own. Fear of the loss of strategic ambiguity and focus on staying on message thus place powerful limits on public knowledge and deliberation during campaign seasons.

A fourth and final factor undermining the public sphere, in Castells's view, is the widespread attention to scandal politics. Castells notes that this is not a phenomenon unique to the United States, as he tracks the destruction of leaders in many countries. Some of his examples include Helmut Kohl's resignation after he admitted that there had been illegal financing of his party, the attempted impeachment of President Clinton for sexual improprieties, and corruption-related scandals involving Spanish and French Socialist administrations. In the absence of any documented incremental trend in corrupt and questionable practices, Castells wonders why the attention and awareness of political scandals have become so salient. He speculates that intense media coverage of scandal is part of a political competition that has elided attention to issues and instead led to political gamesmanship. Castells believes that when political positions are blurred, citizens focus instead on the character of national leaders and party reliability to win votes. Therefore, the best way to win the political game is to bring one's opponent's character into question.

If the public sphere's continuance and health require access to all viewpoints and to information; a nonproprietary, open media environment; unconstrained discussion of major issues; and a viable nation-state committed to protecting the public interest, then current conditions such as those just described put the public sphere in jeopardy. Although it is true that television and traditional media provide some relatively open venues for discussion of public issues,[3] many citizens are dissatisfied with the kind of coverage of social issues and politics provided by corporate media. The prevalence of political fund-raising, opinion polls, privatization of state functions, and scandal politics produce considerable skepticism about how effectively public opinion can actually influence governments and wielders of economic and political power (Calhoun).

Whether new information and communication technologies (ICTs) such as the internet and mobile telephony alone can provide solutions to these problems is highly questionable. This point was recently made by James Carey in an analysis of historical discourses about technological effects and innovations in communication media. He urged caution about the extent to which we can expect new ICTs to move us toward a healthy public sphere and enable a coming together of the polis. Carey revisited the utopian viewpoint on the internet's probable effects that we saw in the 1990s. He noted the failure of its proponents to take into account the social, economic, and political environment in which the internet was developing. He reminded his readers that "all social change is purchased at a price. . . . Securing what we want also entails giving up something we also want. Every gain is simultaneously a loss. Nothing is linear and cumulative" (447). Carey labeled this a pragmatic perspective, and he placed it in a historical context. Despite optimistic predictions concerning the social and political effects of communication technologies such as the telegraph and television, global détente lost ground after World War I and into the twentieth century as media infrastructure and content "faithfully mirrored the politics of [their] time" (449). Since media shifts occur in a larger political, economic, and social context, they have often eventuated in unexpected outcomes.

Keeping Carey's cautionary advice in mind, I want to avoid overly rosy prognostications about the internet's salutary influence on the public sphere. Nonetheless, it is the case that Web-based affordances offer a number of advantages for public discourse that are unavailable in mass media. Among these are affordability, access, opportunities for horizontal communication and interactivity, online forums for discussion and mobilization, networking capacity, and platforms for multimedia. As Matthew Barton observed,

> [t]hough theoretically anyone can build a hobby radio station or even produce films for broadcast on public television, the reality is that hardly anyone possesses the practical means to do so. The dissemination of radio and television content, therefore, is controlled entirely by powerful multinational corporations. The internet, in contrast, allows any computer literate person . . . to create textual and graphical World Wide Web pages. (178)

Although the use of computers and telephony for communication may be less expensive and easier for users than other media, such as print publications or television, critics of new media have questioned how accessible the internet and the World Wide Web are for many groups of people. Those who live in underserved rural areas, who lack the requisite education and skills to effectively use ICTs, or who cannot afford the hardware, software, and connectivity may have comparatively greater difficulty than more advantaged groups in using these new media technologies. The situation is improving, however. As telephony becomes more widespread in developed countries, the infrastructure for wireless communication expands and makes telephony more available to all segments of the population. Kellner points out that "a new generation of wireless communication could enable areas of the world that do not even have electricity to participate in [new forms of] communication and information sharing" (301). Furthermore, the development of public computing, increased library access, and the universal service provisions of the 1996 Telecommunications Act have enabled access to computing for more citizens in the United States (Kranich; Williams and Alkalimat).[4]

Public Communication and New Media Activism

The affordances of the internet and the World Wide Web have enabled a substantial amount of social activism, issue advocacy, and online protest beginning in the 1990s.[5] These developments have inspired a number of scholars to discuss ICTs' potential for encouraging online activism and citizen involvement. Richard Kahn and Douglas Kellner note the internet's potential to be "deployed in a democratic manner by a growing planetary citizenry that is using new media to become informed, inform others, and to construct new social and political realities" (87).

In terms of resistance, the use of the internet in antiglobalization movements has provided dramatic examples of success in focusing public attention on the detrimental effects of exploitive economic policies and the growth of transnational capitalism. An early example of resistance supported by internet-based communications is the Zapatista Movement in Mexico in the mid-1990s. After years of oppression by the Mexican government, Mexican Indians, *mestizos*, and urban intellectuals reacted to the provisions of the North American Free Trade Agreement (NAFTA) that threatened their rights to land and economic security. To avoid victimization and appropriation of their land in the overall process of economic modernization, Zapatista militants set up a structure and initiated preparations for guerrilla warfare. In the course of their struggle, they used telecommunications, videos, and computer-mediated communication to publicize their struggle to the world, force negotiations with the president of Mexico, and advance a number of reasonable claims that had widespread support in Mexican society at large. The Zapatista Movement created an alternative communication network in Mexico and Chiapas and used the internet to link women's groups in Mexico and the United States. By virtue of their media connections and visibility, the Zapatistas were protected from outright oppression and succeeded in increasing public awareness of government corruption, repression, and social exclusion (Castells). The comparative success of the Zapatista Movement and other antiglobalization, internet-based protests, such as the disruption of the World Trade Organization meeting in Seattle in 1999 and the 2000 meeting of the International Monetary Fund/World Bank in Washington, DC, have caused Castells, in particular, to note the importance of a "*networking, decentered form of organization and intervention characteristic of new social movements*" (*The Power* 427; emphasis in the original).

Although Craig Calhoun is more circumspect than some other observers about the internet's potential to serve as a platform for political discourse, he nevertheless notes ICTs' capacity for supporting social justice and resistance movements. He acknowledges that "it is certainly true that the international activism of indigenous peoples, environmentalists, and opponents of the World Trade Organization . . . has been organized in new ways and with greater efficacy because of the internet" (231). While the internet does not, in itself, constitute a public sphere, its potential for point-to-point communication, worldwide access, immediacy, and distribution facilitate offline and online protests and participation by widely distributed groups. Calhoun concludes that "one of the most important potential roles for electronic communication is . . . enhancing public discourse . . . that joins strangers and enables large collectivities to make informed choices about their institutions and their future" (244).

The Effects of Online Public Discourse: Three Examples

In the Habermasian framework, one significant goal of deliberation through public discourse is to affect policy decisions and government action. Therefore, the question arises of whether the internet's capacity to serve as a platform of public discourse has contributed to a reinvigoration of the public sphere. Has discourse on the Web enabled significant social and political action with effects on policy, practices, or social cohesion? Does use of the World Wide Web and other new media by political campaigns, social activists, media reformers, government agencies, nongovernmental organizations (NGOs), and other entities foster citizen involvement, open deliberation, and public participation in policy formation? If so, by what means are these outcomes accomplished? What dimensions and affordances of new media enhance public discourse and decision making? How does online communication facilitate offline social protest and social activism? How does internet-based communication complement and enable use of conventional mainstream media by NGOs, activists, and other groups? Three examples of online social action in crisis management, environmental protest, and political campaigning will illustrate how rhetorical appeals combined with networking and bottom-up communication can promote social cohesion and public involvement in major events and issues impacting the public sphere.

In the aftermath of the attacks on the World Trade Center and the Pentagon in 2001, people initially turned to television and mainstream media for news and images of the events. But in the weeks following September 11, the internet became a rich site for public discourse. A Pew Internet and American Life Project ("One Year Later: September 11 and the Internet") documented and described the ways in which internet-based discourse was used to alert, inform, console, and mobilize audiences. Due to its multimedia capacity, the internet was used to disseminate photographs and user-created video and audio clips. Millions of users viewed images of the damaged buildings, rescue efforts, portraits of victims, and memorial vigils that were unavailable on television.

Charitable organizations and individuals mounted Web sites and sought donations or posted calls for volunteers to assist (Foot and Schneider "Online Structure"). Corporations, government agencies, and individuals posted memorial Web sites to commemorate those who died (Foot, Warnick, and Schneider). Political groups adamantly called for either a violent or a nonviolent response to the attacks. Religious denominations used the Web to provide spiritual and emotional support to their members, and they posted prayers and links to scripture, called for peace, and endeavored to inform and convince users that Islam is a religion of peace and venerates Jesus among its prophets (Schneider et al.). After intensive study of internet-based discursive activity during the 12 months following the attacks, the authors of the Pew Report concluded that "the findings illustrate the importance of the Web as a significant component of the public sphere, enabling coordination, information-sharing, assistance, expression, and advocacy in a crisis situation" (Schneider et al. 25).

Internet-based environmental activism presents a second example of online persuasion and mobilization. An example of this is the work of Greenpeace, an organization originally established in Canada in 1971 that has since become an NGO with worldwide influence. Its international Web site lists links to independent Greenpeace Web sites in over forty countries. Its primary emphases in 2005 were stopping climate change, preventing ocean pollution, protecting forests, resisting genetic engineering, eliminating toxic wastes, abolishing nuclear weapons, and encouraging sustainable trade ("What We Do"). Stories about Greenpeace's internet activity reveal that the organization views the internet as a vital component in its overall media strategy.

In New Zealand in 2003, for example, Greenpeace opposed the lifting of a moratorium on genetic modification (GM) by positioning one of its spokesmen as a knowledgeable media source. Owing to his effectiveness, Greenpeace became, for a time, one of the dominant sources for journalists reporting on the issue.

Television news then directed viewers to the Greenpeace Web site's Free Food Guide and provided the URL (uniform resource locator) when reporting on the story (Morton and Weaver). Another example of Greenpeace's Web-other media connection occurred in June 2000, when delegates to the Oslo-Paris Commission that regulates ocean pollution in the North Atlantic viewed live images of liquid waste laced with radioactivity being discharged into the ocean near a private beach by France's La Hague plutonium reprocessing plant off the Normandy Coast. Greenpeace made the Web cam images available throughout the meeting and on the internet. Subsequently, ministers from 12 of the 14 member countries decided against the practice of radioactive dumping in the ocean (Lortie).

As Morton and Weaver note, "environmental issues are frequently not regarded by media producers as hard news" and that this, "combined with the fact that environmental issues are often very complex in terms of causes and explanations has resulted in low levels of coverage, and reporting what is event-sensitive as opposed to issue-sensitive" (248). Greenpeace thus has to take pains to address its primary issues clearly on its Web site. As an independent, nonprofit organization, Greenpeace must also raise considerable funds through membership subscriptions so as to protect its nonpartisan status (Morton and Weaver).

Greenpeace's international Web site in 2005 (http://www.greenpeace.org/international/) reflected efforts to capture media attention and sustain membership. The site included rankings of electronics firms (e.g., Sony, Apple, IBM) with regard to their policies and practices on disposal of electronic waste. Some of the companies receiving low rankings have subsequently changed company policies on waste disposal of their products. After posting a member poll to identify Latin American leaders who had encouraged and supported deforestation in Latin America, the site exposed those leaders to public embarrassment. The site's article on toxic materials in children's toys contributed to the European Parliament's ban on toxic chemicals in toys. Coordinated use of media events, news coverage, and Web-based discussion forums has enabled Greenpeace to sustain membership funding, hold the public's attention, and influence public policy related to the environment. As Castells observed of Greenpeace, the organization succeeds by virtue of "specific campaigns . . . organized on visible targets [and] spectacular actions geared toward media attention that raise a given issue in public consciousness, thereby forcing companies, governments, and international institutions to take action or face unwanted publicity" (176). Internet-based political discourse plays an important role in Greenpeace's efforts.

Dean for America (http://www.deanforamerica.com/), Howard Dean's 2003–2004 primary campaign for the Democratic Party's nomination for U.S.

president, provides us with a third example of how online resources can be mobilized to support public discourse in the political field. Dean made extensive use of the internet in combination with offline networking to develop a nationwide effort to support his nomination. The campaign, directed by Joe Trippi, Dean's campaign manager, began by using Meetup (http://www.meetup.com/), a Web tool that enables users to identify and bring together people in their geographical vicinity who share their interests. In Dean's case, the Meetup format enabled him and his campaign to bring together groups of Dean supporters to meet face to face. By mid-2000, Dean had 3,000 supporters who had come together via Meetup, and developments escalated from there.

As a result of their meetings, Dean supporters began emailing each other, and some of them posted blogs in his support.[6] By mid-November, the Howard Dean group on Meetup had over 140,000 members, and there were over 800 monthly Dean meetings on the calendar (Wolf). Dean supporters also wrote letters to the editors of their local papers and distributed them through the network so that others could do the same. By September 2003, Dean had raised $25.4 million, mostly made up of small donations over the internet that averaged just under $80 (Wolf).

Dean and Trippi decided to run a bottom-up campaign. They posted blog entries regularly themselves and sought supporters' advice on important campaign decisions, running polls on the site and sometimes following the advice. Supporters responded as positively to the way the campaign was being run as they did to the candidate. Until the third-place finish in the Iowa primary and the incident of the "Dean Scream," the campaign's strategies and use of the internet were decidedly successful. In his review of the campaign's success, Christopher Lydon observed that

> [political pros] have seen the future of progressive organizing, and they know it works. In its small-sum fundraising on the internet, the Dean campaign cleansed the Augean stables of campaign finance when bought politics had come to seem the unbeatable rule. And it put a compelling short list of serious issues on the table for all to argue: the extension of health care, the refinancing and reinvention of public education, responsible realism in a world that wants to respect us. (para. 5)

Dean was the first national candidate to make a sustained run at the nomination through a largely internet-supported infrastructure. The effort evoked a high level of involvement from his supporters and impressive grassroots organizing and fundraising. The implications of this phenomenon for combining online and offline campaign strategies in the future will become apparent when the influence of Dean's campaign on the 2004 Web campaigns of George W. Bush and John Kerry is described in Chapter 4 of this book.

It is these and many other examples of resistance actions and grassroots political organization that have led social theorists Richard Kahn and Douglas Kellner to conclude that

> [b]road-based populist political spectacles have become the norm, thanks to an evolving sense of the way in which the internet may be deployed in a democratic and emancipatory manner by a growing planetary citizenry that is using new media to become informed, to inform others, and to construct new social and political relations. (88)

Since the internet has potentially global reach, it has served to enable worldwide political movements to support social justice and to challenge the forces of transnational capitalism and economic power. It is also worth noting here that the Web and the internet can be used to mobilize groups for ill as well as for good, as Castells's accounts of the online development of the militia movement in the United States (that eventuated in the Oklahoma City Bombing), Japan's *Aum Shinrikyo*, and al-Qaeda have richly illustrated. As I explain in the next section, however, the work of communication scholars on internet-based public discourse has to date been largely critical, whereas I am primarily interested in the use of rhetorical critical methods to study and analyze all forms of discourse, including those with primarily positive results.

Media Shift and the Need for Research of Online Public Discourse

During the past quarter century, a number of media shifts have occurred and are now occurring with increasing frequency (Carey; Herring). Printed texts and continuous television programming generally have given way to ever shorter segments on television and to discontinuous, chunked texts in other media, including print media and the Web.[7] Thus, substantive, integral texts in speech and writing have been displaced, as electronic and Web-based communications are parsed into sound bites, hyperlinked lexias, media clips, and images. As a result, identifying a source or author for a text is becoming difficult or less possible for coproduced or corporate-authored texts. Online audiences and readers of hyperlinked texts select their own point of departure for reading the text and organize it through their reading practices. In many communication environments, writing is no longer unquestionably the dominant mode of communication, having been overtaken by the image (Kress).

Nevertheless, most forms of communication, other than those designed for purely aesthetic or informational purposes, involve some element of persuasion. They thus involve the use of rhetoric, which I take to mean, in Kenneth Burke's definition, "the use of words by human agents to form attitudes or to induce actions in other human agents" (*A Rhetoric* 41). Burke emphasizes that rhetoric is a function of language use rather than a discrete genre of communication, and it is concerned with "the *persuasive* aspects of language, the function of language as *addressed*, as direct or roundabout appeal to real or ideal audiences, without or within" (44; emphasis in the original). An implication of this broad and inclusive conception of rhetoric is that persuasion is not only constituted of words, but also of many forms of symbol use, including images, nonverbal and verbal communication, explicit, sophistic forms of advertising and propaganda, and oratory and public address. Rhetorical forms in online media also include coproduced media discourse, online political campaigns and parody, epideictic discourse in online memorials, and other forms of appeal. Often these are hybrid discourses involving information and aesthetic elements as well as rhetoric, but one of their aims will be more or less explicit appeal to purported audiences in specific communication contexts.

One implication of combining awareness of the acknowledged media shifts described above with a broad Burkean view of rhetoric's function is that the use of rhetoric in new media environments has been understudied by scholars and critics of rhetoric. The need for more research and theory in this area is great, a point that was well expressed by educator Gunther Kress when discussing the pressing need for communication research on new modalities:

> In the domain of representation and communication, the crisis manifests itself at every point: genres are insecure; canonical forms of representation have come into question; the dominant modes of representation of speech and writing are being pushed to the margin and replaced . . . by the mode of image and by others. (17)

Although I would not go so far as to say that there is a "crisis" in the development of theory and critical methods to analyze new media texts, I will agree with Kress that "tools are needed that will allow us to describe what is going on, and theories are needed that can integrate such descriptions into explanatory frameworks" (6). I have already established that online communication can play a substantial role in persuading publics to support candidates, deliberate about issues and policies, and mobilize to resist transnational capitalism and exploitation. In calling for more research on the rhetorical dimensions of online public discourse, I am echoing Zizi Papacharissi's call for "case studies where the internet was used to mobilize support . . . to understand the process through which online discussions can begin to gain political weight" (23). I also concur with John Downey's and Natalie Fenton's injunction to better understand the nature of counterpublic discourse and "the extent to which it knows itself as rhetoric, reinventing the promise of community through discourse" (194).

Communication researchers interested in ICTs have been working in three areas: composition studies, technical communication, and communication studies. To undertake a sustained analysis of public discourse on the Web from a rhetorical perspective, one must first consider some of the work by scholars in each of these areas. I begin by discussing work in composition and technical communication, and then move on to consider relevant work in communication studies. The first four examples were selected with an eye to identifying some theoretical and method-ological problems in studying online discourse. If Kress was right in his observation that established conventions, canons, and genres of expression have substantially changed with the advent of new media, then one would expect such problems to surface, since theoretical frameworks and methods well suited to the study of expos-itory and argumentative prose in print would not necessarily transfer well to digi-tal texts. Thus, the description of rhetorical criticism in the first five examples will help me identify research questions that will be addressed in Chapters 3 through 5 of this book. These chapters will then set out to identify and elaborate the theoret-ical frameworks and methods potentially useful for critical analysis of online rhetoric.

Work on Online Public Discourse in Composition Studies and Technical Rhetoric

Two contributions to *Alternative Rhetorics*—a book described by one reviewer as focusing on research areas that have been "marginalized or ignored" (Ono 461)—focus on rhetorical dimensions of electronic texts. Theresa Enos and Shane Borrowman note that alerting students to the need to examine ethos in the sources they use is a vital part of teaching expository and argumentative writing. They explain how this is accomplished by describing markers of ethos and credibility in three Holocaust denial sites studied by their composition students. The authors of these sites achieved the appearance of credibility by citing credentials and by por-traying themselves as free of bias and concerned with finding the truth. Their pages were well designed, usable, and generally well maintained. Although this is an excel-lent study on the ease of maintaining the appearance of credibility and the diffi-culties of assessing it, it dates from the late 1990s. In Chapter 3 of this book, we see that evaluation of site credibility now is much more complicated than it was before because of corporate authorship and coproduced content. The pervasiveness of Web-based messages with no identifiable author causes recommendations to look to the "credibility of the message source" to be highly problematic.

The second essay in *Alternative Rhetorics* by John B. Killoran provides insights into how disenfranchised authors take up Web authorship and use rhetorical means of self-expression. Drawing on a sample of 110 student home pages, Killoran explored how home page authors used the genre as a vehicle for expression. For the most part, these home pages functioned as sites for carnival and self-commentary (Bakhtin). In particular, the use of parody, genre transgression, colloquialisms, and other forms of irreverence produced texts with a low fidelity with reality and thus "a low affinity between [their] producers and the prevailing system" (129), which devolves from lack of solidarity between that producer and that system—a difference of power. Without a good deal of social capital, the young authors of these sites had little recourse but to resort to skeptical self-commentary. They appropriated discourses of commercial advertising, ludicrous self-promotion, and online parody to fill up their space. To introduce some modicum of interactivity to their sites, they inserted clever pseudo polls to keep readers engaged, filling their sites with forms similar to those they and others had been forced to fill out. As will be seen in Chapters 4 and 5 of this volume, use of interactivity and intertextuality (playing a new text off of an already familiar text) are rhetorical strategies often designed to keep users on sites and involved in their content. Killoran concluded with an apt explanation of parody's online appeal: "Parody inserts [authors'] voices, virus like, into the commerce of mass media, while at the same time redressing their estrangement. It contests otherwise secure precedence and monopoly of institutional agents and practices in the public domain" (141).

While these studies of the online experiences of students and young people focused on their development as critics and producers of online texts, they do reveal ways in which recipients of online persuasive discourse learn to judge the texts they encounter. They also indicate that novice site authors often already recognize the tricks of the trade (such as using Web site stickiness to attract and hold user attention) and are able to intuitively emulate these tricks themselves.

The third study took an innovative tack in studying online activism by comparing and contrasting instances of online social activism across time. In this comparison of three online protests during the 1990s, Laura Gurak and John Logie provided a microgenealogy of internet-based resistance. Drawing on Gurak's earlier work (*Persuasion and Privacy*), the authors described two text-based protests in the early 1990s—protests against the Lotus Corporation's proposed release of Lotus Marketplace and against the U.S. government's planned changes in encryption technology. In both cases, the issues in dispute related to invasion of privacy and security. In the first instance, Lotus Marketplace was a direct mail marketing database on CD-ROM containing personal information on 120 million consumers that Lotus planned to offer for sale. In the second, the proposed Clipper Chip would enable encryption that could be unscrambled by U.S. investigative agencies. The internet protest against Lotus Marketplace included posting of form letters and petitions to be directed at Lotus. Over 30,000 consumers wrote to the company asking that their names be removed from the database. Lotus delayed the database's release and then never published it. The 1994 protest against the Clipper Chip was directed at the U.S. government through emails, Usenet newsgroups, and discussion lists, and provided sample form letters and petitions for users to sign. Despite these efforts, the Clinton administration adopted the Clipper Chip as a federal information-processing standard in February 2004.

The third protest was Web based (not text based) and occurred in the summer of 1999 when Yahoo! acquired GeoCities. Prior to the acquisition, GeoCities had allowed its members free storage space for their Web pages in the expectation that the content produced by users would attract internet traffic and thus revenue. When Yahoo! acquired GeoCities, the company tried to change the Terms of Service agreement to state that all content posted on Yahoo!'s pages was to become the intellectual property of Yahoo!, and that the agreement was to be retroactive immediately. The notice was posted on June 25, 1999. On June 29, the new agreement was brought to the attention of a former GeoCities customer named Jim Townsend. He established a "boycott Yahoo!" protest site that quickly became a central node in a sizable protest against the new agreement by GeoCities homesteaders. Members who were Web designers devised protest banner ads that were widely distributed. Also, thousands of homesteaders took down their pages and posted battleship gray replacement pages, stating their objections to the new policy and including links to Townsend's site. On June 30, *Wired Online* posted a story about the protest with a link to Townsend's site, and the same story was published in the *New York Times* print edition on July 1. By the next day, Townsend's pages had received over a million hits (Gurak and Logie). On July 6, Yahoo! revised its Terms of Service agreement to meet the protesters' conditions, and Townsend then declared the protest over.

This study shows how quickly online protests can attract public attention when they get covered by mainstream corporate media.[8] A posting of a protest site's hyperlink by prominent news outlets can very quickly increase traffic on that site exponentially. When the object of protest is a major corporation, the threat of extensive, negative coverage can effectively cause the corporate entity to modify its behavior. Also, when a community of supporters already exists online, they can be mobilized quickly to apply public pressure in an issue controversy.

As a malleable point-to-point communication platform, the internet enables nearly immediate dissemination of information and calls to action through its distributed networks. This capacity, labeled "speed and reach" by Gurak and Logie, intensifies the internet's usefulness as an organizing tool. The same capacity has proved effective time and again, as in the Zapatista Movement, antiglobalization protests and demonstrations, and in this case, where an online community successfully stood for its rights to its members' intellectual property. Furthermore, as this comparative study showed, the duplicated, redundant texts circulating in the text-based Lotus protest were supplanted in the Web-based Yahoo! protest by a much more rapid and efficient use of the hyperlinked and design capacities of the Web, thus increasing the efficiency and effectiveness of protest action.

The fourth instance of work considered here is a dissertation by Constance Kampf ("Kumeyaay"). In her dissertation, Kampf considered the use of the internet by indigenous peoples to express themselves and expand the reach of their messages. It was shown that their efforts to influence perceptions of themselves had an influence on public policy. Kampf studied the Web presence of the Kumeyaay people, a group of Native Americans residing in California. Her study illustrated how Web site design choices were influenced by rhetoric and culture and how the Web was used as a site of interaction between majority and minority cultures.

In the final study of the five in this group, Matthew D. Barton focused on the uses of interactive and coproduced student discourse, self-expression, and argument. He began by noting that production of Web content has become more difficult since Web page and site construction are becoming ever more technical and complex, using not only graphics and sound but also more programming code. The gap between the requirements of professional Web site development and people's ability to produce Web pages means that "the workers are being gradually, yet effectively, separated from the means of intellectual production" (Barton 178). To begin to compensate for this difficulty, Barton describes three online writing environments that people can learn to use rather easily—blogs, discussion boards, and wikis. He notes that all three are available in "open source" versions and so can be downloaded and installed at no cost. Barton maintains that the nature of the tools we use to express ourselves in public is as important as the content of what is said in determining whether free expression is in fact in play. All three of the writing spaces identified by Barton are simple to use, egalitarian in structure, and designed to encourage users to engage in public discussion.

Blogs are online journals or diaries in which an individual can record personal experiences, thoughts, and opinions. Most blogs focus on a topic or a set of topics and they discuss these on an ongoing basis. Blogs work like discussion boards, but with a few key differences. Unlike discussion boards where the status between participants is equal, blog authors create the site content and may allow users to add comments to the author's posted entries; in other words, the owner of the blog is the principal or sole author on the site. A few blogs are posted by a community of users, however, with several authors responding to each other's work. Of the genres of Web writing, blogs may be the most personal. Barton notes that they are comprised of personal musings and opinions, and he feels that "one of the primary functions of personal blogging is the development of subjectivity" (184), by which, he means a sense of oneself as someone who has valuable opinions, something to say, and preparedness to engage in advocacy and discussion with other people.

Discussion boards resemble listservs or text-only bulletin board systems used prior to the development of the World Wide Web. They allow members to post or respond to "threads" (sequenced communications regarding a topic or issue). Users interested in participating on discussion boards can find a topic of interest, read the prior entries written by others, and post a response. The discussion board software automatically organizes and archives topics and replies. Most boards require users to log in under an identity so everyone can tell who said what, but they are otherwise easily accessible and can enable groups of writers to quickly form a discourse community and engage in rational-critical debate.

Wikis, unlike blogs and discussion boards, offer a radical approach to authorship in that they are Web pages that anyone can edit. Thus, their texts are authored by various contributors editing others' writing rather than by any specific author or authors. To change a node in the wiki, a user can click on the "Edit page" link at the bottom, change the text, and submit the changes. The text that has been entered is converted to HTML by the wiki system. Some wikis allow anyone to modify nodes whereas others allow only registered users. For each node, there is a log of all changes made to it and a record of those changes (Emigh and Herring). One of the most prominent wikis is Wikipedia, a collaboratively authored encyclopedia, at <http://en.wikipedia.org/wiki/ Main_Page>. By enabling constant revision and updating by a number of people, wikis emphasize an open concept to multi-authored texts and thus depart radically from the single-authored text. Barton observes that an exciting use for wikis is online, consultative production of documents such as petitions, resolutions, or consensus-based rules.

Blogs, discussion boards, and wikis, therefore, are writing platforms that enable content development and knowledge production adapted to the nature and purposes of a given communication environment. They serve as sites for persuasion, self-promotion, information dissemination, and other communication functions. They encourage many forms of expression and organization, but rhetorical critics may want to focus primarily on the extent to which they support and further online social and political action and complement other media by providing spaces where citizens can strive for consensus and share a common, community voice.

Work on Online Public Discourse in Communication Studies

Much of the work in communication studies since the mid-1990s that deals with computer-mediated communication has taken the form of ethnographies and discourse analyses of online interpersonal, small group, or organizational communication and text studies of new media using critical or postcolonial perspectives.[9] Rhetorical criticism of positive instances of the use of new media technologies is a noticeably under-researched area.[10] There are at least three possible reasons for this shortfall. First, as I have noted early in the preceding section, persuasive uses of new media to disseminate viewpoints and address audiences do not generally take the form of sustained speech and oratory that rhetoricians have been trained to study. As Dilip Gaonkar noted, "the practice of rhetorical criticism regards oratory, especially political oratory, as the paradigmatic example of public address" (410). He goes on to note that the study of speeches and oratory draws its conceptual orientation from the classical Greco-Roman rhetorical traditions. Because the shift to new media emphasizes discontinuous, segmented texts, media sound bytes, and images, rhetorical critics may be disinclined to apply traditional theories to new media texts. Second, critics witnessing the digital divide that emerged in the early years of the internet and the Web may have viewed ICTs as principally benefiting elites and at the same time constraining the roles played by other, less advantaged groups in the public sphere. Therefore, rhetorical critics may have been at that time disinterested in studying ICT-enabled rhetoric. Third, academics' skepticism in the 1990s about utopian predictions of technology development may have chilled interest in studying its positive effects (Selfe; Tyner; Warnick, *Critical Literacy*). In any case, the two example analyses of online persuasion that I will now describe are critical in nature, but, like those in the preceding section, open up intriguing questions for the study of online persuasion that are pertinent to those to be addressed later in this book.

In a detailed study of 23 Ku Klux Klan-related Web sites, Denise Bostdorff noted a number of dimensions of online communication and rhetoric. Although many of these are not unique to the Web, the medium's distributed, transitory nature lends itself to their development. For example, Bostdorff noticed that "sites from [her] list [of selected sites] frequently had disappeared when access was attempted, or if a site was successfully accessed it often had a new address" (345). While the sudden disappearance of Web sites may be particularly noticeable with hate sites, it also occurs frequently with other noncorporate, nongovernment sites developed by groups for immediate social action. Challenges for researchers seeking to overcome the problems of studying transitory texts will be addressed in Chapter 2 of the present volume.

Bostdorff also noted the medium's ability to create shared identities through reiteration, links to other similar groups, use of Web rings, and other means of encouraging users to visit sites of other mutually supportive groups similar to the one to which they belong. In addition, she expressed concern about how users are to judge the credibility of the online rhetorical appeals they encounter, especially when many users are inclined to evaluate favorably messages that are lively, timely, and relevant to their interests without scrutinizing its veracity and factual basis. As I have noted, problems with online credibility and ethos continue to be troublesome and will be addressed from a theoretical perspective in Chapter 3.

Bostdorff indicated other dimensions of Web-based public discourse that facilitate certain forms of rhetorical action and reception. The anonymity of Web authorship and the authors' distance from their readers may encourage uninhibited negative expression because, as Bostdorff speculates, "rhetors need not fear confrontation over or rejection of their messages" (346). The Web's linking capacity fosters cross-fertilization of ideas by referring users, not only to other sites, but also to messages on other media such as low power radio and public access television. Furthermore, Web sites advertise group specific events in remote geographical locations and encourage users to attend them. (We will see that Indymedia and some political campaigns use much the same strategy in Chapters 3 and 4.) Bostdorff also describes the capacity of the Web to offer tailored content for targeted audiences such as women and children. Her example of the New Order Knights' posting on their Kids' site of the lovable *Sesame Street* character Bert holding a Klan flag and professing to support the Klan is a reminder of how easily images can be selected, copied, altered, and reposted to support any cause.

Bostdorff's study reminds us of how readily amateur (as well as professional) Web authors can use the medium for good or for ill. It also reminds us that the affordances of the internet and the Web can play a significant role in shaping the rhetorical appeal of message content and channeling user response to it.

Fürsich and Robins undertook a textual analysis of the official Web pages of 29 sub-Saharan countries' Web sites. They used a critical perspective to describe the sites' efforts to use the Web to attract foreign investment, tourism, and development funding to their respective countries. What Fürsich and Robins had hoped to find on the sites was "an outlet for vernacular communication (and aesthetics) or an equal voice on the world stage" (203). What they found instead was an effort to "brand" various countries and to reflect an identity mirroring Western interests. The sites accomplished these aims through the use of identity symbols, exotic natural scenes, stereotypical representations of Africans, and appeals to Western readers to visit or invest in the country. Fürsich and Robins also noted significant omissions on a number of the sites of references to domestic unrest, genocide, and civil wars in the host countries.

Their study's approach and method exemplify four of the challenges of Web-oriented research: the difficulty of recording analyzed texts, problems in accessing them, problems providing reader access, and the tendency to over generalize from limited case studies of specific instances to internet-based activity in general. First, while its authors acknowledge that their analyzed texts were used "to indicate an identity in transition at a specific historical moment" (205), they do not specify when that moment was beyond the eight-month collection period in 2001. Second, the authors do not provide specific retrieval information for the quotations that they use from the sites, although they do provide a list of the sites and their home page URLs at the end of their article. Citation protocols require specific page URLs for all quotations from the Web, including their retrieval dates (American Psychological Association; Gibaldi). Precise citation is necessary because rhetorical and textual critics advance claims about the texts they analyze, and readers must be positioned to evaluate these claims by comparing them with the texts described. Third, by failing to provide archival information about the texts, Fürsich and Robins make it difficult, if not impossible, for readers to consult the specific pages they analyze. In describing their method, they say that

> We base our analysis on the poststructural concept of text, but we are also aware that any analysis asks the researcher to "fix" the text under investigation in time and space, even if temporarily. In order to allow for the fleeting material of internet content, we did not download the sites onto our hard drives; rather, we repeatedly accessed our sites via their URL addresses over a period of several months (January to August 2001), during which time some changes to the pages may have occurred. (194)

In early Web research, it was common practice for analysts either to provide precise URLs and citation or to download the sites they analyzed. Precise URLs (to the specific page that is being described) potentially enable readers to consult the site texts in the internet Archive <http://www.archive.org/>, so long as a retrieval date is provided. I say "potentially" because the archive does not store site versions on a daily basis, and there are other problems with citing archived material that will be discussed further in Chapter 2.

Fourth, the authors of this study tend to essentialize the internet as a technology having harmful effects in the public sphere. Despite some caveats on this issue such as the observation that "a significant number of African expatriate Web sites . . . establish and popularize resistance groups against dictatorial regimes" (199), and that countries such as Mozambique and Tanzania sent mixed messages that were not exclusively Western influenced, Fürsich and Robins generally advocate a view of the internet's structure and nature as being hegemonic and pro-Western. They conclude that "our study demonstrates that the aesthetic and logic of the World Wide Web . . . reinforces counterperiphery imbalances of knowledge production" (208) and that "computer-mediated communication . . . does not automatically transcend old categories such as nation and the nation-state . . . but may reinforce them and fill them with new corporate meaning" (208). As I have noted, I believe that critics should not look to the communication platform or its features as per se good or bad when making attributions about how its use affects society. Instead, they should consider how its affordances are shaped and applied by users to address social problems when considering the nature and effects of public discourse in the public sphere.

Conclusion

I began this chapter by noting concerns that have been expressed about the current state of the public sphere in contemporary society. Critics have noted that some of the causes behind its decline include public skepticism about politics and participation in civic affairs and the rise of profit-oriented corporate media. As institutional spaces for open discussion have eroded, many groups and interests have sought out alternative outlets for discussion of major issues facing society. In this environment, the internet has provided an affordable, comparatively accessible platform for spokespersons and community organizers to communicate with their publics. Furthermore, as this chapter has noted, the rise of telephony as well as wireless technologies and mobile computing devices has recently made access possible for many underprivileged groups and citizens in developing countries.

In light of such developments, the use of internet technologies for social activism, resistance, and environmental and antiglobalization protests has been widespread. This chapter described how skillful uses of the internet enabled the Zapatistas and anti-WTO protesters to attract worldwide attention to their causes in the 1990s. These and other examples I have described have indicated how effective online persuasion has been in recruiting movement participants, garnering news coverage, raising funds, furnishing forums for open discussion, providing issue-related perspectives not covered in mainstream media, and enabling the organization of offline activities such as street protests and demonstrations.

In addition, we have seen that the affordability and access provided by the Web have introduced changes in political campaigning in the United States. As the Dean for America campaign in 2004 has shown us, the Web can be a highly useful tool for rapid mobilization and fund-raising by newcomer candidates. This example, like others presented in the chapter, reveals the internet's potential as a platform for public discussion and persuasion and thus for reinvigoration of the public sphere.

Taking a broad definition of rhetorical criticism as study of the persuasive dimensions of communication generally, this chapter has proposed that rhetorical critics take up the project of studying online public discourse. Doing so will require us to rethink and adapt conventional canons of rhetoric and argument analysis. In particular, we need to develop methods for studying discourse that, unlike print and other mediated texts, is often coproduced, interactive, intertextual, ephemeral, immediate, and/or distributed in nature.

The challenges facing rhetorical critics of new media content are many. These critics must begin by reconsidering some of their assumptions about the nature of textual production of speech and print texts. They must generate new methods for critical study of texts that differ in production and form from those they have studied in the past. They will also need to consider issues related to documentation, preservation, and representation of short-lived Web texts, as well as the issue of how to adapt their critical readings to rapidly changing structures and display protocols. One way to approach these challenges is through media-specific analysis, an approach that has been proposed by literary and communication theorists and that will be the focus of Chapter 2.

2

Online Rhetoric: A Medium Theory Approach

The study of rhetoric and the practice of rhetorical criticism historically have been concerned with the persuasive dimensions of discourse. The long history of rhetoric has emphasized the various aspects of the rhetorical process, such as the ways in which speakers use credibility, logic, and emotional appeals to persuade an audience (Aristotle), how speakers can serve as a medium or channel of persuasion (St. Augustine, in Bizzell and Herzberg), how rhetors shape their discourse to suit their imagined audience (Perelman and Olbrechts-Tyteca), or how communicators appeal to others through mutual identification (Burke *A Rhetoric*). Considering these and other variations leads to the conclusion that "rhetoric" is a highly contested term. In this chapter and throughout the book, however, I consider rhetoric in a broad, Burkean sense as "the use of language as a symbolic means of inducing cooperation [or response] in beings that by nature respond to symbols" (*A Rhetoric*, 43). In order to see how a broad, inclusive definition such as this serves us well in studying the rhetorical dimensions of online discourse, we must first consider how more limited definitions of rhetoric might constrain us.

For example, Aristotle defines rhetoric as "an ability in each [particular] case, to see the available means of persuasion" (36). His conception of the rhetorical art emphasized the role of the persuader in using resources to craft a message text. In large part, discourse produced by specific authors or speakers is usually viewed as emanating from a message source.[1] Aristotle's emphasis is still reflected in a focus on the author or speaker's roles as origin of the message. Many forms of online discourse have no single author, however. Is authorship of a major corporate site the product of programmers, Web designers, content authors, automated assembly processes, or other forces? Although corporate authorship is not unique to electronic communication, underlying technologies and current modes of production make using authorial intention as a primary means of judging online message quality problematic.

The dispersion of authorship also does not mean that online messages operate only as a medium or channel of communication. Web site messages are *designed*; they are often carefully wrought attempts to attract and retain audiences. From a Burkean perspective, this may be achieved through various means of identification, such as in-jokes, shared forms of entertainment, user polls, immersive experiences, and Web site customization. When we consider how persuasion occurs differently in online, interactive Web-based environments from its use in context-specific and comparatively more stable circumstances, then a medium-specific set of resources for the rhetorical critical study of online texts seems desirable.

The usefulness of a medium-specific approach is illustrated in N. Katherine Hayles's approach to analyzing digital literature and electronic texts. She has argued that "electronic literature [as compared with print literature] requires a new critical language, one that recognizes the specificity of the digital medium as it is instantiated in the signifying practices of these works" ("Deeper," 373). In her analyses of digital texts, installations, and artists' books, Hayles emphasized the roles of navigating, moving, manipulating, and altering the text as enacted by the user, and she concluded that we "must toss aside the presupposition that the work of creation is separate from the work of production and evaluate the work's quality from an integrated perspective that sees creation and production as inextricably entwined" (373).

In the consumption of many new media forms, there is no "work" to be isolated and read; what users experience is a page created in the moment for a specific time and circumstance. This page may be created on the fly in response to user choice and actions or as an artifact of breaking news or current developments, such that different users are simultaneously seeing and experiencing different texts designed according to their needs and preferences. As David Jay Bolter has observed, "although certain new media forms do share some of the characteristics of mass media, they are not so relentlessly unidirectional" (Bolter 27).

Due to these and other differences between Web-based communication and previous media, an effort is needed to ascertain the significance of these differences in media forms for the theory and practice of rhetorical criticism. In particular, what assumptions about the nature of textual production and the reception of traditional speech and print texts do not hold for texts that are "born digital?" How can existing rhetorical theories be modified and reshaped for the analysis of new media?

The purpose of this chapter is to identify those dimensions of Web-based discourse that make it necessary to adapt rhetorical theories to the study of persuasion on the Web. Specifically, this chapter considers how such characteristics as nonlinearity, differential access, instability, and dispersion of Web texts affect the processes of Web production as well as users' experiences of and responses to Web-based messages. In relation to literature, Hayles has labeled such an approach "media specific analysis," while others have called it "medium theory" (Hayles; Meyrowitz; Ryan). Such an approach attends to the "material apparatus producing the . . . work as a physical artifact" (Hayles 29). With regard to online persuasion, this approach explores medium-specific possibilities and potentialities for persuasive expression.

I want to emphasize here that I am not assuming that the online medium is superior to print, speech, or other media. As will become clear, the Web as a medium for persuasion has many shortcomings. Among these are its instability and the consequent potential for loss of human knowledge, its possible contribution to declines in conventional literacy, and the problem of unequal access.[2] Nevertheless, as I have shown in Chapter 1, the Web is a site for substantial social activism, political persuasion, commercial activity, and various forms of public discourse. It is very important that rhetorical critics produce analyses and commentaries on the rhetorical activities taking place on the Web. To do so, they may draw on existing design principles and develop accounts of new principles shaping online persuasion. In particular, it is necessary to consider how the shapers of online texts exploit the structure and properties of this medium (Herman).

The rest of this chapter charts the attributes of Web-based communication with regard to five elements of the communication process—reception, source, message, time, and space. With regard to reception, I consider how Web content is processed and experienced in ways different from other media—largely because of its non-linear structure and modes of production. This is in contrast to mass media in that Web users are often explicitly and actively involved in the creation of content and meaning. In the Web environment, the message source plays a different role because of dispersed authorship and uncertainty about exactly who the message source is. Furthermore, the form of the message has changed, since messages in Web environments become fragmented because they are so easily dispersed, altered, reshaped, and transposed into new contexts. Time is also a significant factor in how the Web environment differs from other media. Users experience Web sites in time frames impacted by their connection speeds and hardware capacity. Web pages are unstable and short-lived in their born digital form and, in spite of substantial efforts having been made to archive the contents of the Web, they are far from perfect. The work of critics, thus, will be vital in preserving records and commentary on online public discourse. The final section of this chapter considers space—in particular, the implications of the geographical separation between producers and users and the challenges of attracting and holding the attention of users in cyberspace. In each of these sections, the implications of the communicative, temporal, and spatial characteristics of the online environment for rhetorical critics will be considered.

Reception: The User's Experience

Theorists of writing, literature, and aesthetics have explained many senses in which consumption of online digital hyperlinked text differs from reading or listening to other mediated textual forms. For example, scholarly print texts and literary works make different kinds of demands on readers than do texts created for the World Wide Web. Works in print are generally written to be read continuously and in sequence. Authors of such works provide readers with explanations and definitions for what they will say later, and novelists introduce characters and contexts that can be drawn on in later plot development. Public speakers often adopt similar conventions, establishing their character and motives, forming common ground with audience interests, and building arguments for their positions before advancing claims and recommendations. Thus, reader and listener comprehension is viewed as being dependent on sequential development and crafted arguments.

The conventions of reading and processing hypertext and programmed texts such as computer games are shaped by a fundamentally different technology. The consumer of hypertext, for example, chooses his or her own path through the links presented and thus decides on the order in which textual components will be read. The nonsequential reading that results means that online texts generally do not rely on the forms of organization and argument that are characteristic of continuous texts. In these kinds of texts, linear organization and complex argument forms take shape in what McLuhan and Powers call "visual space":

> Visual space structure is an artifact of Western civilization created by Greek phonetic literacy. It is a space perceived by the eyes when separated or abstracted from all other senses. As a construct of the mind, it is continuous. . . . It is also connected (abstract figures with fixed boundaries, linked logically and sequentially but having no visible grounds), homogeneous (uniform everywhere), and static (qualitatively unchangeable). (45)

This characterization of sequential print text and its effects on readers may be somewhat overdrawn. After all, many writers and readers resist this ordered sequence by various means. Writers use digressions, flashbacks, and unconventional text presentations to circumvent it (Hayles, *Writing*). Readers use indexes to spot-read topics of interest; they read texts out of order eclectically. Nonetheless, in print and continuous speech, the order effect is inscribed in the presentation of the text and is something that these readers must work against.

Writers and readers of Web texts anticipate that texts will be read in the sequence chosen by the user. It is for this reason that Farkas and Farkas remind the readers of their Web design text that "the most important difference between print and the Web is non-linearity" (223). Due to nonlinearity, Web designers have adopted many conventions to enable readers to remain oriented and receptive to site content. For example, when a Web page is initially accessed, the reader can see only that part of the page that lies "above the fold," or the limited display area at the top of the page (Lynch and Horton 92). In order to decide where to go next, the user needs to be able to recognize the important internal links, the most crucial information, and the site design (graphic identity) in that space before deciding how to proceed. Writing content for the Web requires an understanding of modular writing that can be read out of sequence and has few content-related prerequisites. Users are best positioned to find their way if they are assisted by search engines or site indexes. These elements compensate for the lack of linear development, but they displace other conventions (e.g., linear organization, sequential plot development, and complex argument forms) typical of scholarly writing, literary narrative, and argument.

Another way of making this distinction between writing and reading print texts and consuming hypertext is to consider Roland Barthes's distinction between the readerly and writerly texts. The readerly text is a finished work and strives for plenitude and completeness. It does not view the reader as a site of production of meaning. The writerly text, in contrast, is incomplete, plural, indeterminate, and it calls upon the reader to supply or fill in meaning. This is in alignment with Barthes's view of texts as distributed in a sort of network. As such, Web texts are produced through corporate authorship, constantly revised, often borrowed, and frequently parasitic on the other texts to which they are linked. In its intertextuality, performative forms, and indeterminacy, the Web text is more like an organism than like a "work." As Barthes noted, on this view "the metaphor of the Text is that of the *network*; if the Text extends itself, it is as the result of a combinatory systemic" (*Image Music Text*, 161). Barthes thus distinguishes between the idea of "text" (always being formulated) and "work" (a finished product).

Nancy Kaplan (2000) has described Web reading, not as deficient, but as *different* from reading print text. Since there is no predetermined next node in the reading process, readers must continually make choices about what to read next. As they read, they proceed by weighing alternatives, constructing forecasts, and then continually modifying their expectations. Kaplan's view is that readers read Web texts with heightened attention, partially because they are constantly making microdecisions about where to go and what to read next while they are reading.

To expand on this idea, we might think of the user as a participant in the creation of meaning in a somewhat different sense than the reader of a book or the audience of a speech. New media theorists characterize the user as a vital element in the creation of meaning and experience because the user creates the text and experiences it as appropriated and altered by means of his or her participation. Two examples might make this point. In their book *Windows and Mirrors*, Bolter and Gromala describe a participatory art installation with two large parallel screens—one featuring projected video and the other a backdrop. A rain of colored letters falls from the top of the screen, and as people pass by, their silhouettes are projected onto the screen while the colored letters settle around them. By changing how they hold their hands and arms, viewers can create different reflections on the screen. Bolter and Gromala conclude that "the experience of this piece comes from the interaction of the viewers with the creators' design. TEXT/RAIN is as much an expression of its viewers as of its creators; it is what the viewers make of it. Without them, the piece is incomplete, for there is nothing on the screen but falling letters" (13).

A second example is offered by Espen Aarseth in his book *Cybertext*. Taking his examples from the gaming culture, Aarseth describes how game players not only work within the algorithms of game designers but also try to change them by introducing new elements that disrupt the game's programming. Through the use of cheat codes and destabilization of the game's programming, players can change the nature of the text itself as they experience it. Aarseth contrasts those actions with the responses of readers who can interpret, conjecture, or withdraw their attention, but who are unable to change the rules or the nature of the game. He concludes that the player's selective movement is "a work of physical construction that the various concepts of 'reading' do not account for" (1).

These two examples from digital art and programmed gaming are apropos here because both of these fields have had a strong influence on the look and feel of the Web. Along with virtual reality modeling and digital cinema, they have produced forms and designs that have influenced Web design and production. The participatory textual creation conveyed in their descriptions can also be seen in the use of the Web. There, users contribute site content by participating in discussion forums, providing comments on blogs, submitting articles to group-authored sites, responding to polls, and authoring book reviews and reviews of other materials. Users who are frequent contributors of such material play an active role in creating the texts that they read.

Of course, the conventions introduced by media shifts do not entirely supplant those that preceded them. The designs of graphic and Web interfaces have relied on the conventions of earlier media since their origin. The window and page as frames for entering into online experience remain with us at present, although they are changing with the advent of portable computing and telephony. Archived articles and many Web pages present a "print" option for capturing their content. Web page links and other materials read from left to right and top to bottom. Web sites are usually parsed into hierarchical layers of greater detail and subordination much like subordination patterns in print texts. Web authors use chunked text, subheadings, italics, and other means of display to facilitate users' reading (Lynch and Horton).

A similar phenomenon can be seen in the representation of visual images. Manovich has shown how the display conventions of twentieth-century cinema and photography are emulated in digital photography and cinematic production. He explained that digital components can be composited into a single object. Their origin in diverse sources created by different people becomes hidden and the result is "a single seamless image, sound, space or scene" (136). He called this effect photorealism—the ability to present composited images so as to make them appear to be faithful photographic or cinematic reproductions of visual reality. This effect is also seen in movies where movements and perspectives make use of the conventions of analog cinema, such as dollying, tracking, and zooming in as part of producing digital images. As Ben McCorkle has pointed out, shifts from one medium to another over time are made possible by conflating one medium with elements of another earlier medium so as to naturalize it in the experience of readers and consumers. This aspect of remediation plays an important role in the acceptance of online content and the ease of use by audiences accustomed to older media forms.

These changes in reading practices have substantial consequences for critics studying rhetorical action in online texts. Rhetorical critics in the twentieth century have in general used source- or text-oriented approaches to analyze public discourse. For example, these approaches frequently have examined the role of the source in crafting arguments and designing persuasive strategies, the text's use of style, the material contexts in which texts are produced, and texts' intertextual relations to other texts. What may be needed for the study of online rhetoric, however, is greater emphasis on reader response to the text because, as I have noted, online texts are often customized for specific audience members. Furthermore, even when customization is not in play, different users construct individuated pathways through hypertexts, assembling different texts as they read (Chen and Macredie). As Chapter 5 of this book indicates, readers can also take up different meanings of intertextually based texts depending on their experience with prior texts and the number of textual allusions they can follow. While the study of textual reception, reader-perceived polysemy, and reader-response criticism has been called for and enacted by rhetorical critics (Ceccarelli; Olson and Olson), the specific situation of the online reader as participant and coconstructor of meaning requires further consideration.

In many Web-based environments, the existence of recorded and archived reader responses in the form of comments on blogs, postings to bulletin boards, or other reader contributions makes possible the grounded study of user reaction. Where explicit user reaction is not available, critics nonetheless can consider the ways in which the Web site text prestructures user response through placement of internal and external hyperlinks, intertextual allusions, opportunities to interact with on-site content, and use of recognizable cultural contexts and intertexts that shape the reader's interpretation.

Reader-response criticism is an approach to the analysis of literary forms developed in the 1970s by Stanley Fish, Wolfgang Iser, and others. Iser, in particular, seemed to anticipate the contemporary reading experience of online users by considering the ways in which texts set expectations for reader participation. In his theory of reader response, he drew upon psycholinguistic theory and phenomenology to describe how readers take up meaning from literary texts.

He noted that a theory of reader reception posits various reader positions. Among them is the "real reader," a person who responded in certain ways at specific historical moments; the "informed reader," a highly competent person capable of observing his or her own reactions (i.e., a critic); and the "intended reader," a persona revealed through the text's anticipation of its readers' norms and values.

Iser views the reader's role as shaped by perspectives represented in the text as well as by the reader's disposition to form a background to a frame of reference for grasping and comprehending what is read. The reader's reception is also shaped by "the repertoire of the text"—the social norms and allusions that make up its referential context (86). The reader is not a passive recipient of textual meaning but instead plays an active role in its creation:

> The reader's task is not simply to accept, but to assemble for himself that which is to be accepted. The manner in which he assembles it is dictated by the continual switching of perspectives during the time-flow of his reading, and this, in turn, provides a theme-and-horizon structure which enables him gradually to take over the author's unfamiliar view of the world on terms laid down by the author. (97)

Iser thus echoes Barthes's distinction between the "work" and the "text" by describing the reading process as "a dynamic *interaction* between text and reader" (107;

emphasis in the original). Furthermore, he describes the reader as a textual wanderer navigating the text whose viewpoint is shaped by the repertoire of the text and his or her own dispositions and perspectives. The text must be gauged to the reader's capacities, avoiding overly complex and also overly familiar structures, for "over-strain and boredom represent two poles of tolerance, and in either case the reader is likely to opt out of the game" (108).

For Iser, the reading process takes the form of an emergent actualization of a set of potentialities embodied in the text and in the reader's faculties of perceiving and processing. The text offers pathways, structures, expectations, and frames of reference. The reader, in response to the text, creates meaning by exploring some possibilities (and not others) in light of his or her own experiences. When considering intertextuality in Chapter 5, I show how this process is made possible by potentialities embedded in the text itself.

Due to its conception of the text as a relatively open system that can be appropriated in various ways, Iser's approach provides a means for critics to consider how readers might read and react to different texts in a Web environment. By considering how the textual structuration anticipates user response, reader-centered criticism can consider how the text prestructures its readers' experience and also incorporates their active role in the coproduction of meaning.

Source: The Rise of the Authorless Message

In recent times, it has become a convention to attribute credibility to messages based in large part on the expertise and reputation of the message source. This practice, however, is largely an artifact of the modern period. In his genealogy of ethos in *The Encyclopedia of Rhetoric*, James S. Baumlin (2001) traced ways in which the conceptualization of ethos or credibility has historically been adapted from and applied to its social contexts. He noted that in Aristotle's *Rhetoric*, ethos was thought of as a form of artistic proof—what was said in the speech to make the speaker *appear* to be worthy of credence. During the medieval period and the Renaissance, however, many works were written in the service of the church or other institutions and were anonymous. What was important was the content of the message rather than who wrote it. During the subsequent modern period, disinterest in the author was supplanted by an increasing tendency to think of the work as the author's product. Baumlin explained that the rise of capitalism and private ownership led to concerns about copyright, plagiarism, and ownership of texts, and preoccupation with the identity of the author once again emerged as a predominate concern.

It is useful to keep in mind the extent to which attributions of ethos are based on a sign relationship. Sign relations move from something that is more apparent, visible, or accessible to conditions that are comparatively less observable. For the jurors of ancient Greece, what was observable was the speaker before them: what he said, how he spoke, and how he represented himself. What could not be observed but only surmised were his intentions, motivations, and perhaps prior actions. In contrast and for contemporary audiences, what has been accessible and presumably documented are the author's or speaker's credentials—education, publications, professional experience, and the like. From these supposedly material facts, audiences and readers draw conclusions based on what is presumably "outside the text."

As I establish in Chapter 3 on message credibility, practices on the Web indicate that the close connection between the perceived expertise and reputation of a message source and users' readiness to trust source credibility in gauging the quality and accuracy of message content seems to have loosened. There are many reasons for this development. First, many sites do not contain information about the source of site content. As a frequently cited 2002 study has shown, a substantial proportion of public Web sites (between 35 percent for some site types and 75 percent for others) failed to provide information about who was behind the site; more than one-third provided no address or phone number, and one quarter provided no information about who owned the site (Consumers International). Another study of comments by users on Web sites they visited indicated that users did not rely primarily on site authorship as an index of credibility. Instead, the most frequently mentioned characteristics (in decreasing frequency) were design look, information structure, information focus, company motive, and information usefulness. The study's authors speculated that one reason for the emphasis on the look of the site may be the speed at which users generally make judgments. They noted that "the visual design may be the first test of a site's credibility. If it fails on this criterion, Web users are likely to abandon the site and seek other sources of information and services" (Fogg, Soohoo, and Danielson).

A second reason for the declining importance of the message source is that the Web as a communication environment may not foster writing and production practices that support reliable identification of sources to establish message credibility. This change may be due to the fact that communicative practices in a particular mediated environment are shaped by the structures of the environment itself. Thus, "technical specifications predispose users toward certain communicative choices, social dynamics, and normative outcomes, which in turn enable them to realize their situationally-grounded goals" (DeSanctis and Poole, 1994, cited in Emigh and Herring 1). A substantial proportion of the sites on the public Web are produced by institutes, corporations, interest groups, nongovernmental organizations, and other entities, and large corporate sites are produced by teams of people, including content experts, technical writers, graphic designers, and technical experts (Lynch and Horton). When they are called up, many of these sites are assembled through automated processes designed to customize site content for the individual user at a specific point in time. This process was well described in relation to an article in a newspaper series:

> When a user accesses the first page of *Trippeldrapet,* the text "content" is pulled from a database and instantly coded in code tailor-made for the user's browser make and version, with banner ads chosen for her geographical area or even demographic or interest. Thus, the elements that make the page the user encounters are not stored in one place and may never show their face again. Even when the page is recalled from an archive, much of it will have changed. (Fagerjord 321)

Thus, the role of the author as message source is displaced by group authorship, automated assembly, and reliance on databases elements such as generic templates and scripts (Manovich). The text that the user reads is a product generated by Web designers, browser settings, and automated page assembly processes containing content customized for the reader's needs.

Third, the advent of coproduced sites, such as discussion forums, blogs, and wikis, has caused the group authorship phenomenon to become even more pronounced. These platforms for producing online content were introduced and explained in Chapter 1, but it is worth reviewing the coproduction process for each of them here. Discussion forums take shape as registered users contribute content by communicating with each other, and they are structured by their moderators and software that automatically organizes and archives topics and replies. An example of such a site is Slashdot (http://slashdot.org). As I noted in Chapter 1, blogs often function as online personal diaries or sites of commentary and are usually authored by a single author or a small group of authors whose point of view dominates the site. Comments from readers are frequently allowed on blogs, but they can only be viewed if visitors click on a link to open them. Wikis are group communication mechanisms that allow users to create and edit Web page content freely using any Web browser. All wiki users are potential authors and editors (Emigh and Herring). The most famous and frequently used wiki is Wikipedia. All of these platforms allow user-created content to a greater or lesser extent, and in many cases the users are anonymous. Although blogs usually have specific authors whose expertise and reputation may stand behind the site, wikis and many discussion forums do not have identifiable authors. To judge the credibility of these sites, some other standards for evaluating their ethos are necessary.

Three factors, then, have contributed to a declining emphasis on source credentials and reputation to judge the credibility of Web site content. These include the absence of authorship information, the emphasis on corporate authorship, and an increasing reliance on coproduction to generate site content. As Chapter 3 of this book argues, it may be advisable to rethink the modernist conception of credibility as reliance on the message source and to shift to a framework for judging the credibility of messages that arise in the context of a distributed field of production oriented to specific purposes, functions, and values. The means for evaluating message credibility in this framework would grow out of the context in which online content is produced.

The Versatility of the Web and the Fragmentation of the Message

As Manovich has noted, a significant component of new media texts in their digital form is modularity. The Web text, for example, is made up of modules—HTML code, images, media clips, chunks of text, hyperlinks, and other components. The modularity of Web texts has in turn been influenced by other coexisting media forms.

For example, in tracking changes and points of convergence in newspaper, televised, and Web-based news over time, Lynne Cooke documented progressive changes in the appearance and functionality of news presentations during a period when formats for presenting news content converged and complemented one another. Her study of front-page print news, news Web site home pages, and televised news spanned the latter part of the twentieth century and extended into the early 2000s. In the 1970s, news presentation moved to a cleaner, more streamlined and scannable presentation that was followed in the 1980s by more condensed reportage in the use of space as well as the introduction of multiple points of entry for reader consumption of news content. When the internet joined the media mix in the 1990s, however, use of thumbnail-sized images, visually scannable graphics, banners, and font display capitalized on multimedia appeal to condense content presentation and consumption. This trend migrated to televised news through use of the segmented screen divided into a scrolling ticker tape presentation at the bottom, accompanied by a display resembling a dropdown menu on the left or right, and the remaining main content panel featuring the anchor or a visual of a breaking news story. Cooke noted that this kind of multimodal display has "been popular with the public who are said to appreciate the 'more news at a glance' approach" (41).

The modularity of digital messages' texts and images furthermore means that each of these components can be taken up, modified, and recontextualized. For example, through the use of Photoshop, a one-dollar bill can be altered to sport the face of George W. Bush instead of George Washington's on the currency of the "United States of Fraud," or the audio on the president's State of the Union address can be digitally altered to contain statements that were never in the original address (Wilkins). As Bolter and Gromala observed,

> [v]arious digital forms, and particularly the World Wide Web, borrow from, reconfigure, and remix all of the principal media of the twentieth century (film, radio, audio recording, and television), as well as two great earlier media: photography and print itself. (90–91)

This process is also greatly facilitated by the existence of sizable libraries of individual elements such as that found in Google Images. These libraries provide large archives of elements that can be readily appropriated and altered. It is also enabled by the absence of controls on digital alteration of images. As David Perlmutter observed, "a picture, once made, is no longer a controllable entity—by copyright or by holding it up to some nonexistent negative. It is malleable, a *tabula rasa* for Photoshop, but also for anyone to make any point" (62). Digital alteration is not unique to new media; producers historically have made use of others' materials and made modifications to them in various ways. The internet, however, makes this process very easy and has contributed to a fundamental change in the way various works are produced.

This remediation of media content has led to fragmentation of message texts as content is dispersed into many media forms and variations. Although remediation may disrupt the unity and coherence of the orderly text, it nonetheless makes possible forms of political parody, satire, and commentary not seen in prior media. Since "the closest thing we have to a converged medium is in fact the World Wide Web" (Bolter and Gromala 98), we therefore have a venue that can combine lyrics, music, visual animation, and other art forms in imaginative ways. As I show in Chapter 5, users encountering such material are well positioned to reflect on the contradictions, absurdities, and implications that can so readily be found in the contemporary public sphere.

Time: Web Text as Ephemeral

Time is a significant dimension along which new media differ from print. First, Web users do not constitute a mass audience since each of them consumes Web content at his or her own pace and within the constraints of the user's environment. In contrast, audience members watching a national newscast or entertainment program can be considered to be a mass audience. The programming and advertisements they consume are designed to appeal to them in the aggregate and according to general demographic characteristics of age, income, and interests. Content is "pushed" out to them, meeting the needs of some viewers and not others. Web users, on the other hand, initiate online experiences in their own time and as individuals, although they may form communities of common interest related to mass media content (e.g., Bury).

Second, Web browsing experiences of users are dependent on such factors as bandwidth capacity of their connections, CPU speed and memory size of their computers, software they have installed, and other factors. Constraints on these elements of users' computing environments impact the speed and ease with which they can access and enjoy various forms of Web content. A student in a recently wired university computer lab with high bandwidth networking will experience a high-end multimedia Web site very differently from a person living in a rural area with a modem connection and a slow operating system. Time, then, is a vector that affects how Web-based texts are received and enjoyed.

Third, impediments and difficulties in archiving Web content and the development of technologies underlying Web site production make preservation problematic. The rapidly changing new media environment that has seen the development of HTML coded pages, Java, Flash animation, media clips, streaming media, portable computing devices, mobile telephony, and other new technologies has led to conditions in which some elements of online content have a short life span and, given the current state of archiving, cannot be preserved. Furthermore, the preservation of digital photographs is also compromised by the nature of storage media. Julianne H. Newton has noted that the lack of physical storage on film causes real difficulties. She notes that "archivists struggle to protect their digital records, yet longevity estimates for a compact disk range from three to seven years, requiring repeated transfer of digital files to new media" (8).

To address the problem of losing material that is "born digital," there have been widespread efforts to establish archives designed to preserve digital content in the European Union as a whole, and in the countries of France, Germany, Austria, Australia, and Sweden, which are particularly directed at preserving materials in their official government sites and those registered under a specific country's domain name (Edwards). In addition, the Internet Archive (IA) was founded in 1996 as a nonprofit organization by Brewster Kahle to store and provide access to digital content. One of its collections is a general set of Web pages (now accessible through the

Wayback Machine interface at http://www.waybackmachine.org/). As of January 2004, its holdings contained more than 30 billion Web pages (Edwards).

The preservation enabled by this resource is only partial, however, for a number of reasons. First, any medium that is used to store data is potentially vulnerable to accidents and natural disasters. The Web pages are stored on servers in various geographic locations and, while the IA collections are stored at multiple sites, there is still the risk of loss because of possible destruction of the physical facilities where the servers are located and because of the sporadic nature of the frequency of updates. Second, storage media (tapes) can degrade to the point where their content becomes irretrievable. Third, software applications and data formats become obsolete ("About the Internet"). While some Web site construction protocols such as cascading style sheets (CSSs) and PHP programming have been successfully captured and stored in the archive, others such as JavaScript have not. Images stored on a server different from that of the originating host often do not load in archived pages. Adding to these problems is the fact that the IA can only store materials that exist on the public or visible Web, excluding proprietary sites such as newspapers, databases, subscription-based journals, and other sites that require registration or fee payment for access.

The ephemeral and partial preservation of Web content and the limitations of existing archives hold many implications for rhetorical critics of online discourse. Since the work of textual critics depends in part on their readers' access to the materials they analyze, durable links must be provided to the texts that are studied, or researchers must save the content themselves and make it available to their readers after obtaining copyright permission from the content creators. While this process may seem on its face to be no different from what is required for other forms of textual criticism, it nonetheless presents unique challenges. The archived documents may no longer function as they did for their original audiences because of changes in software applications and data formats. Other shifts in technology and storage platforms also may develop so as to prevent a critic's readers from accessing the texts in their original form. Since Web sites are dynamic artifacts, the opportunities to reproduce them may decline with the passage of time.

Difficulties in archiving Web contents present a second set of implications for critics of online discourse and intensify the importance of their work. Through their descriptions of the texts they study, commentators and rhetorical critics can recapture and preserve situated public discourse at the time of its original production. When cross-site links, multimedia downloads, and other dynamic content can no longer be reproduced, critics' accounts of the historical situation, rhetors' motivations, and audience effects can furnish a relatively enduring record of the nature and functions of public persuasion in its context. As they have in the past, rhetorical critics can thus provide a record of rhetorical action and its historical effects.

Space: The User as Spatial Wanderer

When commentators on communication technology development first began predicting what the internet would make possible for users and the public, spatial metaphors suggested visions of a future with unlimited opportunities for human habitation in virtual space. Howard Rheingold's groundbreaking 1993 book, *The Virtual Community: Homesteading on the Electronic Frontier*, described an online group of people that grew from hundreds into thousands of residents between 1985 and 1993. Images of discussions, celebrations, and common activities in this "space" fill his book. From that time and as electronic communities and the World Wide Web developed, the internet medium has been known as a space—cyberspace, the information superhighway—where users participate in Multi-User Dungeons (MUDs), post their Web sites in domains, participate in online gaming and other forms of communication at a distance made possible by new ICTs. Early utopian characterizations of this "space" portrayed its potential to become a level playing field where everyone could openly participate free of the embodiment and material constraints of physical space.

More recently, there has been an increasing awareness that the space of cyberspace is not unlimited. As I noted when discussing the role of time in technology access, the size of a computer's memory and one's location in physical space experientially affect how ICT-based communications occur. For example, one's distance from the service provider infrastructure, proximity to other users, and equipment quality all affect the nature of online experience and users' reception of content. Furthermore, to connect online and produce Web-based content means that a Web author must have a service provider, storage *space*, and a Web site *address*.

Design factors also cause concerns about the limits on communication space. Commercial firms and advertisers are aware that entry pages must load very quickly and that the limited display area above the fold on a site's home page is valuable real estate. For this reason, Web design manuals admonish site producers to establish limits on file sizes (Farkas and Farkas; Lynch and Horton). To be effective, a page must be viewable without scrolling to read the text. More recently with the development of small laptops and portable communication devices, the relationship between display capacity and content quality has become even more a matter of concern.

The asynchronicity of internet-based communications and the variety of communication devices also mean that one's audiences are not gathered in space at a given point in time to receive the messages directed at them. While organizations and Web authors do collect information about user behavior, attitudes, and demographics, the information is imprecise. The absence of immediate physical presence (such as is available in response to a speech where the audience is copresent) means that Web authors must work from attributions about users' attitudes, values, and beliefs. They must proceed on the presumption that the audience they imagine will resemble the one they are addressing and that their appeals will resonate with the users' interests. If not, their users constantly have the option of leaving their site and going elsewhere. As Manovich has observed, "if there is a new rhetoric or aesthetic possible here, it may have less to do with the ordering of time by a writer or an orator, and more to do with spatial wandering" (78). Elsewhere, he compared the anonymous observer navigating virtual space with the flâneur—a person moving through space, mentally recording and immediately erasing what he encounters. Manovich mentions that this "navigable space . . . is a subjective space, its architecture responding to the subject's movement and emotion" (269). Thus, the habitation of cyberspace is often transient and self-fulfilling.

The presence and use of space, both in the online environment and in its material situatedness, are significant for rhetorical analyses of online texts. Web authors cannot assume that readers will remain engaged with what they read, and so authors must expend a good deal of effort just to sustain their users' interests. A reading environment comprised of hyperlinks, navigation bars, discourse chunks, multimedia clips, moving characters, and other elements must be so designed that these elements have continuity and meaning for the user. In Chapter 5, I explore some of the strategies that Web authors use to influence user reaction and to craft a coherent, although often oblique, message. These include use of clever visual design, parody, pastiche, and allusion that appeal to what the user "always already" knows and thus can convert the online experience into a puzzle or game that keeps him or her involved (Warnick, *Critical Literacy*).

Conclusion

That the form of communication changes when there are shifts in the media environment is a well-recognized fact. Walter J. Ong traced the work of Ramus, a sixteenth-century educator who developed pedagogical theories during the shift from oral to print culture and literacy (*Ramus*). Ramus produced grand display schematics that "proceeded by cold-blooded definitions and divisions, until every last particle of the subject had been dissected and disposed of" (Ong, *Orality*, 134). At the time, Ramus's visual displays and diagrams were remarkably well suited to dissemination in the new medium of print.

Even while changes in media shape content, newly emerging media forms retain many conventions of the media that preceded them. This process can be seen in film production during the 1990s. Lev Manovich has traced the rise of digitized (computerized) visual media, but he noted the myriad ways in which filmmakers made use of new production technologies to represent special effects, movements, and textures characteristic of conventional filmic representation. In a new media age, we continue to rely on the practice of viewing video images through a frame, and every effort is made in digital production to retain the illusion of conventional analog film and photorealistic representation.

In part, the retention of conventional display protocols enables new media forms to become increasingly naturalized as media consumers become acclimated to them. McCorkle noted that during the age of the rise of print culture in the seventeenth and eighteenth centuries, print as an interface became invisible and at the same time influenced the shape of discourses in other media. In a similar way during the past decade, the Web has become naturalized in our media experience. Its advent portends changes in the way we think, work, and write, but these changes have been made possible by a media form that retains conventions of prior media while introducing new ones of its own.

Due to this mix of the old and the new, rhetorical criticism of online persuasion can retain the categories and critical methods used for analyzing oral and print discourse and at the same time incorporate new critical methods. By considering how online users read and process Web texts, critics can analyze site design and structure. For example, site appearance and usability play a crucial role in whether users will remain on a site long enough to read its contents and be influenced by its appeals. A Web designer's primary aim is to entice users to remain on a site and to explore its contents (Burnett and Marshall; Manovich). Since users' first impressions are based on the look of the page, visual design may have become as important as verbal style. In considering the effectiveness of Web-based persuasion, rhetorical critics might explore the extent to which the goals of site producers are being met through site design and visual appeal.

Furthermore, critics of Web discourse might consider how on-site communication hails or interpellates its users (Althusser). What values does the site assume its readers to hold? How are its users positioned and addressed? What signs are there that site authors know their audiences—the texts or art forms they consume, their levels of knowledge, and how they identify themselves? On sites that include user-contributed reactions and responses, criticism can focus on explicit reader responses so as to ascertain whether a site was successful in shaping its design, structure, images, and verbal appeals so as to appeal to its expected audiences.

In addition to verbal style, visual design, and site structure, critics can also examine the processes of invention in different ways from those used in print texts. Although it is unusual to find extensive, complex arguments online (because of the nonlinearity of Web texts), we do see other means of engaging users cognitively to sustain their interest. For example, interactivity, described by Burnett and Marshall as "the golden fleece of technical perfectionism" (76), is ubiquitous online. How are feedback mechanisms embedded in the site text so as to include users in the process of establishing what they read and see in site content? As will be noted in Chapter 4, interactivity may occur between the user and site content (e.g., contributing content), the user and the medium (e.g., clicking on hyperlinks, customizing site display), and users and other users. How has the site been planned to include opportunities for interaction with users? What forms of coproduction are available for user-created site content?

A second form of invention is the use of intertextuality, a form of interreference among texts in which an already familiar text is invoked or played upon in a new textual context (Warnick, *Critical Literacy*).[3] Its common vernacular form is parody that intentionally copies the style, organization, or other features of a familiar text or situation by way of humorous imitation. Web sites exist in a larger cultural and textual context, and site authors can draw on this context through allusion, parody, pastiche, and other forms of humor. The intertext in which the site exists thus becomes a resource for content development, and adroit use of this context is a form of invention in much the same way that the use of common topics served as a reservoir for argument in Aristotelian rhetoric.

A comparable form of intertextuality can be seen when it is used in a mass media environment. In considering the function of intertextuality in television programs and music videos, Fiske noted that the theory of intertextuality "proposes that any one text is necessarily read in relationship to others and that a range of textual knowledges is brought to bear upon it" (108). Noting that intertextuality exists "in the space *between* texts" (108, emphasis in original), Fiske aligned it with general cross-reference in genre, character, shared convention, and ideology. He distinguished horizontal intertextuality between explicitly linked televised primary texts from vertical intertextuality between those texts and other secondary texts such as criticism and publicity about the primary texts.

Fiske's theory can contribute to our understanding of Web-based intertextuality in certain ways. It emphasizes the significance of genre in intertextual cross-reference. Genre functions to control and shape the polysemic potential of representation in both televised and online visual media. The viewer's recognition of a message as parody, mystery, adventure narrative, or some other form provides the viewer with a frame for interpreting that message. Fiske's insistence that intertextuality is "inescapable" (115) grows out of the realization that all texts, be they electronic, print, or oral, are understood primarily by their relationship to other texts. Instead of being fixed in a universal empirical reality, meanings emerge from interpretations of socially and historically situated viewers.

There are nonetheless noticeable differences between the uses and dissemination of intertextual references in late-twentieth-century television viewing and Web-based intertextuality in the early twenty-first century. In the earlier period when television content in the United States was produced by three major networks, television audiences were divided into fewer segments than they are now with the rise of cable television and digital transmission. It was easier to exploit the preferences and prior knowledge of a mass audience that tuned into *Dallas* and *Cagney and Lacey* every week. Television viewing was more structured and systemically programmed, and television shows were situated within a stable intertextual environment by virtue of their scheduling and program placement. Their writers and designers could readily draw upon viewers' immediate experiences with the programs that preceded and followed them, as well as their shared knowledge of current events.

In contrast, the current media environment includes dispersed, disaggregated media audiences. It has been destabilized by proliferation of media devices, eclectic audience preferences, and development of new media technologies that enable audiences to consume media content that is targeted, if not personalized. Intertextuality continues to function to control the polysemy of media messages, but it does so by exploiting polysemic potential. As will be seen in Chapter 5, this exploitation takes the form online of first, identifying a wide range of polysemic interpretations and second, combining them into online multimedia presentations interpretable by a wide range of audiences in a wide range of ways.

The remaining chapters in this book use a case study approach to illustrate how adaptation of existing rhetorical theories and use of current theory can be applied to the analysis of rhetorical dimensions of online public discourse. Chapter 3 argues for a field-dependent model for analyzing message credibility in a prominent international Web-based alternative news site. Chapter 4 explores the forms and effects of interactivity in two Web-based political campaigns. Chapter 5 describes how multimodal intertextuality is deployed by parodists and critics of the contemporary public sphere. My hope is that the case study approach will illustrate how heuristic theoretical frameworks for studying online rhetorical activity can emerge from the analysis of specific instances of online public discourse.

3

The Field Dependency of Online Credibility

Many Web-based venues for public discourse present content designed to influence publics on consequential matters. Examples of such venues discussed in this book include alternative media sites, political campaign sites, and sites of political commentary. Discourse on these sites often makes explicit or implicit arguments concerning public policy, human welfare, social justice, government corruption, and other matters of social import. Public discourse sites seek to raise awareness or incite action on the part of users. The claims they advance are knowledge claims, and they often take the form of arguments supported by evidence and reasoning. The questions for users then become "What counts as knowledge?" and "How do I evaluate it?"

In the era of the dominance of print media and readily identifiable authorship, such questions were generally comparatively easy to answer; one looked to the credentials of the message source or the entity producing the message to evaluate its credibility. As this chapter shows, however, that avenue of assessment is now often difficult to pursue. Nevertheless, producing and evaluating argument is an epistemological process and a significant component of knowledge production. Since the coproduced, distributed communication environment of the Web presents some challenging questions about message credibility, this chapter considers what is to be done about evaluating online messages and proposes a potentially useful framework for judging them. The latter part of this chapter then provides a case study of credibility on the global alternative media site, Indymedia, to illustrate the usefulness of the chapter's proposed framework. To set the context for this inquiry, it might be useful to begin by considering how conceptions of message credibility have changed over time.

The Author, Ethos, and Source Credibility in Rhetoric

Prior to the growth of the World Wide Web and other new media forms, there was consensus among communication researchers and information specialists in the United States about the nature and functions of credibility. The credibility of a message was judged primarily according to attributes of the message source, especially expertise, reputation, believability, and trustworthiness (e.g., Boyd; O'Keefe; Schlein). The presumption was that messages were products of an identifiable entity and were to be judged, in part at least, according to the credentials of their producers. As I have observed in Chapters 1 and 2, this way of proceeding is frequently problematic when applied to online environments because messages there are often coproduced by many individuals whose identities may or may not be known, and also because a substantial proportion of existing Web-based messages do not include information about the message source (Consumers International).

Considering a different model for online credibility that is not source oriented might seem reasonable when we realize that the current tie between credibility and authorship only came into prominence during the modern period. In the history of rhetorical theories, this particular view of credibility was preceded by other views that grew out of various cultures in which persuasion was practiced. In Athens in the fourth century BCE, for example, ethos was tied to the speaker's character as portrayed in the speech itself (ethos) rather than to any external knowledge of the speaker's position and education (source credibility). George Kennedy equated Aristotelian ethos with "the speaker's success in conveying to the audience the perception that he or she can be trusted" (*Aristotle, On Rhetoric*, iv). Audience perceptions were based on how the speaker constructed a view of himself as possessing such traits as courage, self-control, prudence, and liberality that were in accord with the values of the Athenian culture. Kennedy took pains to make a distinction between this view of ethos and our own concept of credibility when he said that "Aristotle . . . does not include in rhetorical ethos the authority that a speaker may possess due to his position in government or society, previous actions, reputation for wisdom, or anything except what is actually contained in the speech and the character it reveals" (38). Athenian audiences thus developed their view of the speaker by listening to what he (or his speechwriters) said about his character and actions.

As a point of contrast, consider the notion of ethos as discussed by Hugh Blair in his *Lectures on Rhetoric and Belles Lettres* in eighteenth-century Scotland. Blair described the credible speaker as someone who understood the standards of taste of his time, was knowledgeable about the matters on which he spoke, and possessed the necessary education and refinement to construct speeches that were seemly, well styled, and suited to the tastes of the listeners. Blair maintained in Lecture XXVII that "no one should ever rise to speak in public without forming to himself a just and strict idea of what suits his own age and character, what suits the subject, the hearers, the place, the occasion: and adjusting the whole train and manner of his speaking on this idea" (Golden 104). Like Aristotle's view of ethos, Blair's was centered on audience perceptions of the speaker as he spoke, but instead of explicit, adroit self-representation, the speaker was to be judged according to his taste and education as reflected in his performance. Speaking occurred in a milieu where "aesthetic sensitivity was viewed as a sign of social competence and education" (Warnick, *The Sixth Canon*, 129). Thus, it was essential for speakers to adapt to the situation by showing good judgment and a sense of good taste in everything they did as they spoke.

Aristotle's and Blair's theories of ethos illustrate the idea that, prior to the eighteenth century, notions of ethos were embedded in the cultural and social mores of host societies. They remind us of the possibility that modernist theories of credibility also may be culture dependent. It is important to keep in mind James S. Baumlin's argument that the work as a product of a specific author is an artifact of the modern period. Preoccupation with the status and expertise of the author has thus moved us away from the idea of ethos as a form of artistic proof in the text and toward the idea of source credibility as an external authorizing mechanism for judging the veracity of what is found in the text.

However, has our recent reliance on author credentials and expertise only been yet another passing manifestation of source credibility? Might we consider Michel Foucault's prediction that

> as our society changes, at the very moment when it is in the process of changing, the author function will disappear, and in such a manner that fiction and its polysemous texts will once again function according to another mode, but still with a system of constraint—one which will no longer be the author, but which will have to be determined or, perhaps, experienced. (Rabinow 119)

It may be that Foucault's prediction is coming true, and more quickly than one might have supposed. It also seems that his prediction may apply especially to a vital form of public discourse—writings and content addressed to users on the World Wide Web.

The author's role may be receding in importance because, in the absence of a stable print or material face-to-face context, users are placed in the position of making attributions from a variety of textual cues rather than reaching conclusions based on what is connected to an author's credentials and known reputation. One theory of online credibility was suggested by Nicholas Burbules who observed that the need for selectivity and the expectation of speed and efficiency in online searches may lead users to make judgments differently than they do in other mediated environments. He noted that the Web operates like a self-sustaining reference system and that the processes by which users make judgments in that environment are often circular. A user discovers a site authored by an academic group and follows link paths to find out more about them, or finds a claim on one Web page and does a keyword search to find out more about it, or follows link pathways on the site, or examines links to the site to discover more about the link context of which it is a part. Since the internet and the Web are comprised of distributed, networked systems, Burbules observes that users make use of "distributed credibility." That is, they rely on a number of factors that, taken together, enable them to make field-dependent, comparative judgments about the presumed credibility of a source. The processes they use are not unlike those used by the jurors of Aristotle's day who compared what speakers said to other texts, cultural values, and the seemingly probable in order to decide what to think of the messages they heard.

The move to view Web-based texts as distributed systems and as representations is further supported by Roland Barthes's (1977) distinction between the "work" and the "text." He observed that viewing a text as a "work"—converting it into an object, as it were—is problematic. Barthes notes that a work's explanation "is always sought in the man or woman who produced it, as if it were always in the end . . . the voice of a single person, that *author* confiding in us" (143; emphasis in original). Barthes instead views literary texts as a multidimensional space in which a variety of writings, none of them original, blend and clash. Text on the Web is usually not a centered artifact but instead a decentered "tissue of quotations drawn from innumerable centers of culture" (146), and Barthes emphasizes that, seen in this way, texts continually imitate gestures that are always anterior and never original. Like Barthes's text, those found on the Web are usually produced through corporate authorship, continually revised, often borrowed, and frequently parasitic on other texts to which they are linked. Barthes claims that "the metaphor of the Text is that of the *network*; if the Text extends itself, it is as the result of a combinatory systemic" (161).

The implications of this view of the Web text as immersive and to some extent distributed should be pondered. That users often do not judge Web sites as "works" and do not look to authorship as a form of legitimation has been established in empirical studies.[1] Whereas the idea of ethos as attribution based on a confluence of factors is not new and emerges in other media environments, the internet as a distributed system lends itself particularly well to the notion of the Web text as decentered and networked.

To return to Foucault's prediction that, upon the disappearance of the author function, some other mode of constraint might take its place, we might begin to think about what that could be. In addition, we might further extend our thinking about how credibility functions in online environments. When the conventional signs of credibility to which we are so accustomed are absent, to what other signs do people have recourse? When what we have before us is a text that changes constantly, presents an opaque interface to the common user, and operates in a multimediated, hypertextual context, what we bring to bear in judging it may well rely on a host of factors emerging from a larger system. These include what other sites link to the site in question, whether its content is supported by other content in the knowledge system, whether its stated motives coincide with the presumed effects of its use, how well the site functions, and whether it compares favorably with other sites in the same genre.

The remainder of this chapter proposes an alternative model for thinking about how ethos or credibility operates in Web-based persuasive communication. This framework is suggested by Stephen Toulmin's model of field dependence. Many readers will be familiar with this idea, which has been widely applied to the study of arguments. In his 1969 book, *The Uses of Argument*, Toulmin indicted the idea of rationality as then exemplified in formal, syllogistic forms of reasoning that are field independent, or context independent. He offered instead a model of argument that was ordered and reasonable but at the same time situated in fields of practice and knowledge production. The notion of "field" in his model suggested an epistemological context in which arguments took shape and which implied the criteria by which they were to be judged. In his theory, Toulmin defined the concept of "field" operationally; he said that two or more arguments can be said to belong to the same field when they draw on the same principles, procedures, and protocols. For example, arguments in the field of law rely on evidence introduced in the case, probable reasoning, and relevance to precedent, whereas arguments in the field of medicine rely on patient symptoms, research protocols, prior clinical trials or treatment outcomes, and scientific findings. On this view, the credibility of an argument is evaluated according to standards indigenous to the field in which the argument is made. Similarly, one can describe user reaction to the credibility of online sources according to the field or context in which the site is located. That is, users may judge sites according to the procedures, content quality and usefulness, functionality, and values and norms important in the field in which the online site operates.

To establish the usefulness of a field-dependent model for judging Web site credibility, I begin by reviewing research findings about how users actually judge Web site content. Then I elaborate the ways in which a field-dependent model comparable to Toulmin's can help us consider how context plays a role in credibility judgments. To illustrate this process I provide a case study illustration showing how a specific international alternative media site (http://www.indymedia.org) establishes credibility with its users even though most of its contributors are anonymous and its content is comprised of stories by hundreds of writers who submit their articles to an open publishing link on the site.

What the writers and readers on this site share is a set of values related to social justice, preservation of the environment, resistance to globalization, and commitments to egalitarianism and open process. These shared values and modes of operation function to enhance the credibility of persuasive messages and arguments posted to the site. Unlike mainstream news content that is vetted through a centered editorial process, Indymedia's site content is based on collaborative coproduction, so its users must apply standards different from those applied to large-scale mainstream media news outlets and some other alternative media sources. Due to the anonymity of its authorship and the wide range of contributors to the site, this site is particularly suitable for establishing the usefulness of a field dependent model for judging credibility based on the context in which its articles are produced and consumed. The appropriateness of such a model is worth looking into because conventional standards for making credibility judgments used in judging print and other mass media are problematic when they are applied to the Web, as shown in the research results reported in the next section.

Research Findings on How Users Judge Web Site Credibility

At this point, then, it is necessary to consider how users do in fact appear to make judgments about the ethos or credibility of the content they access on the Web. Fortunately, there has been a good deal of research on this question, and the results indicate that people make use of a variety of means, depending on context and their needs, to decide about the credibility of Web sites. After an extensive review of this research, Wathen and Burkell developed an account of the processes people use for credibility judgments in Web environments.

They claim that, upon entering a site, a user makes some immediate, preliminary judgments based on its surface characteristics. These include whether the site has a professional appearance, whether the interface is easy to use, whether the site is well structured, and other factors related to the site's appearance and functionality. If the site fails on these criteria, the user is likely to leave the site. The second level of evaluation is related to site content. Here, users often look for some information about the message source, and if that is lacking or indeterminate, they look to other factors. These include relevance to their needs as well as the recency and accuracy of the factual information reported on the site. If the site passes muster on these criteria, then the user might move to consider whether its content is congruent with his or her knowledge of the topic and easily applicable to his or her immediate needs (Wathen and Burkell). This staged model of credibility judgment indicates that users in a modular, hypertext environment rely on a confluence of factors to make considered judgments about the credibility of online content. This account nevertheless omits what seems to be an important dimension of credibility on the Web because it does not explicitly consider the senses in which Web-based credibility judgments rely on context-specific cues and criteria. That is, its account is generic rather than sensitive to the specific epistemological fields in which such judgments are being made.[2]

Another survey of user behavior dealt with this problem by considering the context-specific nature of credibility judgments. Rieh studied users who were seeking out, respectively, travel information, information about purchase of a new computer, material in research databases, and medical information. These users prioritized the credibility indicators on these sites very differently, depending on the context and purpose of their searches. For the travel and computer tasks, users focused on topical interests and affective aspects, but when they sought out medical information they were much more concerned with the site's cognitive authority. As might be expected, they made judgments based on graphics and information structure more readily when considering travel and computer purchases than they did when looking for medical information and research materials. The implication of this comparison is that field-independent models accounting for users' credibility judgments may not be adequate to explain how credibility is judged on the Web.

Revisiting the Fogg, Soohoo, and Danielson study briefly discussed in Chapter 2 in which respondents to 10 different categories of Web sites listed the criteria they used to evaluate Web site credibility, we find that nearly half of them reported using visual cues, while the next most frequently used indicators, in the order of priority, were information design and structure, information focus, and scope of the site. The identity of the author or site operated was ranked eleventh out of eighteen aspects in importance for judging credibility. If a field-dependent, context-based model is used, it is likely that some combination of factors related to site appearance and content will come into play, and that how those are combined will depend on the field in which credibility judgments are being made.

A Field-Dependent Model for Judging Web Site Credibility

The remainder of this chapter describes a field-dependent model for evaluating the credibility of persuasive messages on the Web and illustrates its usefulness through application to a case study. It is important to think of this as a modular framework, in the sense that various standards of judgment applied by users will grow out of the epistemological context in which persuasive appeals are made. This proposed framework works in much the same way as Toulmin's model of argument that has been used to analyze arguments. His model included four field-dependent components—evidence, warrant, claim, and backing—and has been widely accepted by analysts of argument structure and function (e.g., Inch et al.; Prosise et al.; Toulmin et al.). The components of the model include evidence (accepted facts or premises), warrant (the reasoning link between data and claim), claim (the conclusion or resolution proposed), and backing (general theories, accepted protocols, precedents, and principles supporting the warrant).[3]

Toulmin noted that the nature of these components will vary with the field in which an argument is made, but that in all cases, the arguer moves from a starting point or evidence to a claim by means of a reasoning link supported by backing, for that is the very nature of argument as a form of speech. Toulmin's argument model, then, looks like Figure 3.1.

It is important to note here that Toulmin's approach to argument seeks to dissociate it from formal logic and other contexts in which the level of surety about

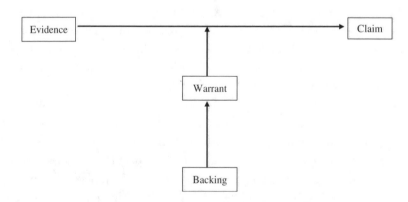

Figure 3.1. Four Basic Parts of the Toulmin Model of Argument.

the claims being made is certain or approaches certainty. He was primarily concerned about discourse in practical contexts. He compared his approach to argument with that of "generalised jurisprudence" (7), an enterprise in which claims are put forward, disputed, and determined according to rules and procedures that have "hardened into institutions" (8). In Toulmin's model, the locus of field dependency is to be found in the backing for the warrant (or reasoning) that is used to support the connection between the argument's starting point, or evidence, and the claim that it makes. Toulmin noted that backing is often tacit and only made explicit when a warrant is challenged. For example, a juror may claim that "the testimony and evidence introduced in this trial is insufficient to convict the accused because it is inconclusive." When challenged on this point, the juror may remind the others that it is necessary to find the accused "guilty beyond a reasonable doubt." This field-dependent standard of proof is particular to the field of criminal law but has epistemological status in the context of jury deliberation.

Toulmin noted that backing in different fields, when used to support knowledge claims or recommendations, may have very different field-dependent characteristics depending on the field. It may, for example, take the form of taxonomic classification, legal principle, statutory stipulation, or moral or ethical tenets. Indeed, in their chapter on ethics as a field of reasoning, Toulmin, Rieke, and Janik noted that, although some people may view ethics as outside the scope of epistemology and rationality, values nonetheless play a significant role in our decisions. The questions may always be raised, "Was our action *justifiable?*" "Were our views on this issue *well placed* and *appropriate?*" They conclude that "once these questions have been stated, the whole machinery of rational criticism and practical argument [and the epistemological status of our claims] may be called into play" (416). The idea that accepted principles and values can play a role in knowledge formation in some fields is an important one to keep in mind when we contemplate the status of credibility in fields such as ethics and law that rely on accepted principles and standards for backing. Some of the specific examples and the case study described later in this chapter clarify how field-specific standards for judging Web site credibility may vary from field to field yet remain stable within specific fields when Web site credibility is evaluated.

Now, let us consider how Toulmin's model might be adapted to a process for analyzing Web site credibility within fields or contexts. As I noted in Chapter 2, attributions of credibility are based on a sign relation. Sign relations move from something that is more apparent, visible, or accessible to conditions that are comparatively less observable. For example, people evaluating message credibility often consider features of the message itself as signs of the accuracy, reliability, comprehensiveness, and validity of the content of the message (Rieh). Absence of errors in the system text, inclusion of important information relevant to the topic, and provision of information consistent with the users' prior knowledge would in some contexts operate as "evidence" supporting attributions of credibility to the site, and the user's judgment about the site's credibility would be comparable to a "claim." The "warrant," or reasoning link, moves from observable conditions (what is provided on the site) to an attribution. As I have noted, the crucial component of the model is "backing." Toulmin notes that backing functions to support the links made in the reasoning process, so it plays a crucial background role in each context. As Toulmin explained,

> [s]tanding behind our warrants . . . there will normally be other assurances, without which the warrants themselves would possess neither authority nor currency—these other things we may refer to as the *backing*. . . . The kind of backing we must point to if we are to establish [the warrant's] authority will change greatly as we moved from one field of argument to another. (103–104)

What I show in this chapter is that the role of field-related principles, practices, protocols, precedents and other assumed knowledge-generating elements in judging Web site credibility is significant. Therefore, it is not feasible to insist on field-independent standards, such as the expertise and reputation of the source, for judging Web sites generally. Let us consider, then, how Toulmin's model for argument can be adopted for analysis of the credibility of online persuasive messages.

The model for assessing site credibility for a genre of sites providing ratings and reviews for movies provides my first illustration of field dependency. An examination of the Internet Movie Database (IMDb) reveals a number of visible features on the site (evidence). These include fast loading times, smooth functionality in viewing movie clips, a comprehensive database of movie titles, and numerous movie reviews. On the other hand, the site provides little information about the credentials of its reviewers, has no contact information except for an email address, and has a lax privacy policy. Based on these characteristics, users might conclude that the site is generally credible, but that its recommendations need to be cross-checked with other sources and that one should be cautious about registering on the site because of concerns about how it might use registrants' information (conclusion). Users might reach this conclusion with some reservations because the site functions so well and so rapidly to meet their needs. The field-dependent aspect of this judgment comes through clearly, however, when one looks to the backing for the warrant. For the genre or context of reference sites in the entertainment field,

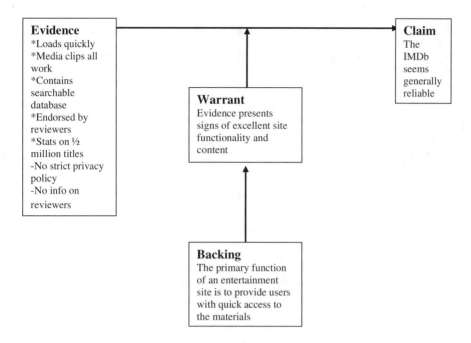

Figure 3.2. Field-Dependent Credibility of the IMDb Site.

the most important function of a Web site of this nature is to provide quick access to information about entertainment content (backing). As shown in Figure 3.2, the site succeeds on this measure.

If we consider analysis of credibility in a field of knowledge other than entertainment, we may discover that the warrant and backing function very differently. Suppose a user who has difficulty breathing, a bad headache, fatigue, and a cough decides to use a search engine to look at health Web sites for information on her condition and discovers a number of sites. On one of them—Medline Plus (*MedlinePlus*)—she finds external links to 28 sites containing information on sinusitis as well as links to medical journal articles with discussions of symptoms, causes, diagnoses, and treatments for this condition. She also notes that Medline Plus is linked to the National Institutes of Health and is located in the .gov domain. After thoroughly reading the information on the site, comparing it with other sites in the same genre, and realizing that, of the 10 symptoms for chronic sinusitis on the site, she has all 10, she decides that the site is highly credible. Her assessment of the site's credibility using Toulmin's framework can be charted as shown in Figure 3.3.

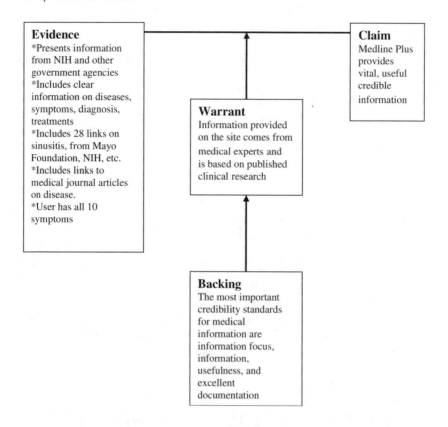

Figure 3.3. Field-Dependent Credibility on the Medline Plus Site.

Comparison of credibility on the entertainment site and the health information site reveals the extent to which the epistemological field and context of a site affect how its credibility is evaluated. When they make judgments about the content of the messages they encounter, users bring to bear the standards arising from the epistemological and experiential fields in which they are working. For the entertainment sites, the stakes involved in making a poor decision based on the site's recommendations are relatively low, as compared with sites containing medical advice and information. People using entertainment sites are primarily interested in finding what they want quickly, and so they appreciate site features such as rapid searches, "top films" listings, show times and ticket listings, and high functionality for running media clips. In contrast, user criteria for medical sites are strikingly different, probably because the accuracy of information provided is so important. For these sites, users emphasize the quality of information, corroboration from medical research, and relevance to the users' needs and concerns.[4]

The influence of backing, then, comes through most clearly when users draw upon a number of factors and prioritize them according to the underlying assumptions and principles inherent in the field of evaluation. The relationship between the "evidence" (the signs found on the site) and the "claims" (their decisions about its credibility) is governed by the relevance of field conditions to the decision at hand. For the entertainment site, users could disregard the uncertainty about who authored site contents and the integrity of the site's privacy policy in favor of convenience and ready provision of the information they wanted. For the medical information site, however, knowledge of where the information was obtained, how thoroughly it had been vetted, and its congruence with the user's own experience of her illness outweighed ease of navigation, aesthetics, convenience, and other factors. It is also important to note that credibility judgments lie along a gradient from high credibility to having low credibility. They are always made in light of the user's prior experience and knowledge, the immediacy of the user's need, and a number of other factors. The extent to which the field of judgment plays a role in the perception of message credibility in online persuasion will now be shown through an extensive case study of the global alternative independent media site, <http://www.indymedia.org>. Rather than being controlled or vetted editorially, the Indymedia site's content is based on collaborative coproduction, so its users must apply standards different from those applied to large-scale mainstream media news outlets and some other alternative media sources.

Indymedia: Alternative News and Distributed Credibility

The global Indymedia site is the central node in a worldwide network of online independent media centers (IMCs), and it is run by a committee of volunteers from these centers. Its history has been well rehearsed in the literature (Beckerman; Gibson and Kelly; Kidd; Pickard). The first local Indymedia site was started in Seattle in 1999 just prior to the WTO protests in that city. Its intent was to enable independent journalists and media producers to produce and distribute media coverage of antiglobalization protests. Indymedia centers use open source software and provide a platform where users and contributors can send and receive information. Since 1999, Indymedia has expanded from a U.S.-centric orientation to one with worldwide coverage by seeding and encouraging over 150 local IMCs in Africa, Latin America, Oceania, and Asia, as well as the global north.

Unlike coverage in corporate-owned, mainstream media where "a protest's likelihood of making the news is determined by how well it fits into issues already in the news" (Owens and Palmer 337), Indymedia's aim is to cover stories and news perspectives elided in the mainstream news. Although it shares this aim with other alternative media sites, it has a number of characteristics that make it unique among alternative media sites. First, it is very prominent among alternative Web-based media. As Atton noted, "the global network of Independent Media Centres . . . is arguably the most extensive and visible radical media internet project serving the broad international coalition that comprises the contemporary anti-capitalist movement" ("Indymedia," 147). The number of visits received by Indymedia centers is very hard to determine precisely because many centers do not keep track of their server traffic. Opel and Templin went to some lengths to obtain this information for site traffic on Indymedia local servers in 2003; they found that visits to IMC servers in Alberta, Baltimore, Boston, Idaho, Utah, and Vancouver, when combined, ranged from approximately 90,000 in January to over 250,000 in March. (This was during the period prior to the start of the Second Iraq War.) A 2006 statement regarding traffic on the Indymedia site estimated that Indymedia as a whole has between 500,000 and 2 million page views a day ("Indymedia's").

Second, Indymedia affords its users the opportunity to readily publish stories to its newswire by following easy instructions and clicking on the "publish" link. It therefore enacts the values of free speech and inclusiveness by exhorting its users to "become the media" by submitting stories and comments about the content that they find on the site. Third, its egalitarian nature is shown in Indymedia's organizational structure and the transparency of its procedures. Work is done by committee; decisions are consensus based and are reached after listserve-based open discussions (Beckerman). Fourth, Indymedia enacts principles and practices that appear to be altruistic, insofar as possible. It has an open copyright policy, operates on donated equipment and expertise, and consistently supports social justice issues. These values, principles, and practices are compatible with Indymedia's alternative media field, and they also underlie the characteristics of its form of credibility with its users.

Indymedia's reliance on flattened hierarchies, wikis, temporary group representation, and collectivism also distributes prepublication responsibilities for vetting and judging its content among stakeholders and Indymedia users. As one email to the general listserv noted: "The burden of maintaining quality [with wikis] is higher than with a normal web site, but the opportunity for equal participation increases the number of eyeballs and keyboards attending to the task at hand" (cited in Pickard 28).

The Indymedia links on its Web pages indicate the topics emphasized on the site. They include support of labor advocacy, the environment, social justice,

TABLE 3.1. Topic Coverage in Issues 69, 70, 72, and 74 of the Global Indymedia Site

Topic	Issue #69	Issue #70	Issue #72	Issue #74	Total
Labor advocacy	4	2	1	3	10
Environmental protection	1	0	1	1	3
Environmental justice	1	3	0	1	5
Government actions against protestors	4	6	5	4	19
Social justice	1	0	0	3	4
Antiglobalization	3	2	6	3	14
Peace protests	1	2	0	1	4
Immigration issues	1	1	2	2	6
Student protests	0	1	0	1	2
Human rights	1	1	0	0	2
Other	3	2	5	1	11

Note: These totals are for the articles in the features column only for the named issues.

antiglobalization, and human rights. As Table 3.1 indicates, a survey of 80 stories in the features column on the global Indymedia site during 2005 showed an emphasis on labor advocacy, coverage of repression against activists and protesters, and antiglobalization protests.

These stories originated in 28 countries in North and South America, Europe, Asia, Oceania, and the Middle East. Although precise user information is not available (Atton and Couldry), scholars who write on alternative media have numbered among Indymedia's audiences the dispossessed, immigrants, trade unionists, students, working people, anarchists, and the unemployed (Atton, *Alternative*; Downing).

Authorship as Authority

It is important to note that users who seek out information about author credentials and editorial authority on the site are unlikely to find such a strategy to be very helpful. The site prides itself on the fact that it "has no central office . . . and no address, phone number, or fax" ("Indymedia's"). The organization's mantra—"Don't hate the media, become the media"—is enacted in its open publishing system and minimal editorial control of its content. Since contributors can readily use open publishing software to post their stories directly to the site's newswire, anyone with immediate access to and knowledge of newsworthy events can contribute content. The advantage of this system is that immediate, first person accounts, photographs, and commentary produced at the sites of protests and other events can be placed on the site. The disadvantage is that legitimation of editorial review does not stand behind them. As one observer noted, "readers accustomed to this model understand the disadvantages of open publishing; they realize that any media—open or not—should be read with a critical eye" (Ballve, para. 15).

Onsite content is displayed in three columns: one main column with featured stories in the middle and two side columns (language options, alerts, internal site links, and other IMC links on the left; events notices and the newswire on the right). The featured stories in the central column are selected from Indymedia's open newswire by a committee of volunteers. An April 2006 sample of center-column features in issues 75 and 76 of the global Indymedia site revealed that, of 40 articles, 33 were anonymous, 6 were indeterminate (using only first name), and 1 was identifiable.

Authors are anonymous for various reasons. Indymedia is a collaborative, nonhierarchical effort; most volunteers work as a team, and individuals are not inclined to identify themselves. Furthermore, many of the newswire authors could be singled out and persecuted if their identities were to be disclosed. Atton has observed that alternative media include among their authors social activists, "native reporters" who bear witness to what is happening on their home turf and movement intellectuals who seek to place dissident activity in a wider social context. He concluded that, among these groups, "the frequent use of anonymity and pseudonymity . . . suggests an aversion to the professionalization of intellectual activity based on personality and reputation" (Atton, *Alternative*, 120). In short, these writers appear to have little motivation to reveal their identity.

Open publishing advocates argue that in such an environment, "notions of legitimacy and credibility that go hand-in-hand with the tradition of journalism are disregarded in preference for a free dissemination of information" (Gibson and Kelly 10). They also maintain that the anonymity of sources and lack of strict editorial control make it possible to communicate viewpoints and information that otherwise would not be accessible. For these reasons, focus on credibility of the message using standard measures of source credibility does not seem to be a highly salient factor for Indymedia users.

The question now remains, how might these users and other visitors to the site use alternative standards to judge the credibility of Indymedia's content? To illustrate the usefulness of a field-dependent approach to assessment of online message credibility, I now describe some of the mechanisms enabling users to judge Indymedia's credibility, as well as the kinds of backing on the site that make attributions of credibility possible within a field-dependent framework.

The Field Dependence of Message Credibility on the Indymedia Site

One way to think about how credibility judgments are made on a site such as Indymedia is to compare its epistemological context with that of conventional, corporate-owned mainstream media. Historically, national newspapers, network-owned national television news, and cable news channels have subscribed to a set of principles and values that presumably ensure that their coverage is dispassionate, objective, editorially reviewed, and based on research using third party informants such as subject matter experts or government officials (Bennett; Entman). These measures are said to ensure that the news that readers see and read is factual, professionally produced, and committed to furthering the broad public interest by providing reliable information backed by institutional authority.

In its "about" link and in answer to "frequently asked questions" (FAQs), Indymedia makes it clear that its commitments and ways of proceeding contrast in nearly every way with those of conventional media. In their account of the site's history, Indymedia's authors emphasize the use of a "democratic open publishing system" and the existence of a "decentralized autonomous network" ("About Indymedia"). They claim to "work out of a love and inspiration for people who continue to work for a better world, despite corporate media's distortions and unwillingness to cover the efforts to free humanity" ("About Indymedia"). It becomes immediately apparent, then, that the backing for credibility assessment on the site will not rely on centralized editorial control or review, nor does the site's commitment to create "radical, accurate, and passionate tellings of the truth" reflect an intention to rely on objective reporting or on official sources for news content.

Instead, site authors maintain that those who report news events on their site are working with local IMCs that "have explicit policies to strongly deter reporters from participating in direct actions while reporting for Indymedia" ("Indymedia's"). In addition to using reports from writers whose aim is to provide first-hand and complete descriptions of available information, the site's FAQ maintains that the site does not represent special interests but instead works to open up discussion of important issues by all interested parties. While there are editorial guidelines posted to Indymedia's "publish" link, there is no designated editorial collective that edits articles. Although there is a "Newswire Working Group" that "clears the newswire of duplicate posts, commercial messages, and other posts that don't fit with Indymedia's editorial guidelines" ("Indymedia's"), all other stories that are submitted are posted. Those writers who want to submit featured articles must make use of

Indymedia's existing format, links internal to its network, and photographs to accompany their feature. A committee then selects the features to be included in each issue's center column. From these procedures, we can conclude that, insofar as possible, Indymedia is committed to an open publishing system with very minimal oversight of the content its writers post to the site.

To those who wonder how or whether they can believe the news they read on Indymedia, the site's FAQ page responds that purely objective reporting is not possible because "all reporters have their own biases; governments and massive for-profit corporations that own media entities have their own biases as well, and often impose their views on their reporters" ("Indymedia's"). It urges its readers to look at all reports they read on the site with a critical eye, and notes that writers who publish to the site invite "public debate about their positions from any readers of the site." As will be seen when we discuss the "comments" sections that follow each story, readers' views from across the political spectrum can be voiced, and the factor that works to back credibility on the site is its commitment to "provide a safe space" for the voicing of all viewpoints. Thus, the reader must look to his or her own assessments, based on what is said on the site, in its comments, its external links, and on other sites in order to decide whether the content on the Indymedia site is to be viewed as credible. It is the diversity of views and the lack of editorial control that work to position the reader in this way.

Furthermore, as the site explicitly states, it is opposed to "the increasing corporatization of society and culture" ("Indymedia's"). Its operations are, therefore, funded by donations made by users on the site and by other noncorporate interests, and the site is maintained wholly by volunteers. Thus, a major principle to which the site subscribes is that grassroots coverage of events must grow out of mechanisms that are not profit driven or proprietary. Due to this commitment to a nonproprietary, nonprofit approach, Indymedia uses open source software and donated server space and equipment. It provides all of its original content "free for reprint and rebroadcast, on the net and elsewhere, for noncommercial use, unless otherwise noted by the author" ("Indymedia's"). Its disavowal of influence by special interests and the absence of a profit motive may contribute to its readers' inclination to attribute some measure of credibility to the Indymedia site because the site's content is not controlled by any corporate interests.

A final factor in the production of the site that potentially contributes to Indymedia's credibility is that it is a decentralized network. As I've previously noted, the global Indymedia site has no physical street address or location. Furthermore, all the decision making is distributed. In addition to the committee (with rotating membership) that selects articles for the features column, users can contact the site working group via an email address, or register with working groups via a Web interface. It should be noted that the global Indymedia site is a coordinating node in an international distributed network of IMCs from which most of its content originates. Most of the site's decision processes are decentralized and work through lists where interested parties have signed on. Deliberations also occur online using internet relay chat where people communicate in real time. These highly distributed decision-making processes preserve the egalitarian structure of Indymedia and prevent special interests from controlling its content. For users who distrust centralized, corporate management practices, such an arrangement may be reassuring.

Credibility Judgments as Reflected in Users' Comments

A final place one can look for indications of the field specificity of content on the Indymedia site is comments posted by users in the open comment space that follows all stories. There, one finds discussions of story content and significance, as well as disagreements about the veracity of what is reported. As Toulmin noted, backing is most likely to be made explicit when the warrants and tacit support for a statement's conclusions are challenged. Therefore, the content of the posts challenging what is reported is a promising place to look for explicit, field-dependent statements of backing. To illustrate how this occurs, I consider one signal instance of such an exchange.

An article posted on September 26, 2005, claimed that Ojeda Rios, a former leader of the armed struggle for liberation of Puerto Rico, was killed when the FBI assaulted his home in Puerto Rico. After he was shot, the FBI officials did not enter the home and refused entry to family, supporters, and others, allowing Rios to bleed to death from his wound. There was a demonstration by Puerto Ricans in New York City to protest the incident, and Rios's funeral in Puerto Rico was attended by many Puerto Rican dignitaries. On its face, then, the position stated by the original post and some added information would result in an argument like Figure 3.4.

The backing for this argument immediately became contested in the comments following the Rios story, as some respondents introduced other facts not reported in the original post, and others questioned whether the story was credible (see comments included in "United States"). The story's credibility began to be challenged in the fourth comment when a commenter asked whether there is "any evidence that our comrade was assassinated? . . . But you can't blame everything on the U.S.A. or go around accusing it of things without some evidence. I would welcome more factual information in this article. . . . If this is true, it needs to be

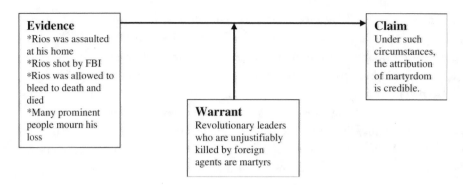

Figure 3.4. Field-Dependent Credibility on the Indymedia.org site.

SUBSTANTIATED." The sixth comment claimed that "the FBI was willing to pay $500,000 for information." The seventh comment asked whether Rios was "the same person on the FBI's most wanted list" and noted that he was a bank robber of a federal bank. The ninth comment claimed that Rios was "a wanted thug." The eleventh commenter pleaded for further proof "or at least a reasonable suspicion that the US did this." The thirteenth commenter asked, "who is the editor of this website? I'm totally perplexed as to why you would print something like this. I mean if you have any evidence it would be a great story. But otherwise you're just completely undermining your credibility." The final comment in English went as follows:

> I wish I could know more about the situation in Puerto Rico and it's [sic] history with the U.S. I would like to believe our brothers [in] PR in relation to the assignation [sic] of their leader, but the point of this website and indymedia is to find out what is the truth for yourself. I would like to know what other people know about the issue. I'm not going to ask anyone what should [I] think but I will ask everyone what they think and I'll decide what I think. ("United States")

This series of comments in the open comment section following the Rios story indicates how credibility operates for Indymedia authors and readers. The fourth and eleventh comments call for further evidence and substantiation of the story's reportage, whereas the thirteenth comment calls for some form of editorial responsibility to review or screen what is reported. The criteria invoked by these three responses assume that the same sort of objective journalistic standards seen in mainstream news should also be applied to Indymedia content.

The final comment, however, reminds readers that such a field-dependent standard is inappropriate when applied to Indymedia. Instead of relying prima facie on established news accounts, the users on Indymedia are supposed to "find out what is the truth for yourself." To do this, they can turn to the story content, information about the event in external links and other sources, and their own prior experience and knowledge. A core principle of Indymedia is that users should explore all views and then decide what they think for themselves. Remarks by the final commenter rely on that principle, which serves as backing to explain how credibility judgment should work on Indymedia. This user's statement is supported by Indymedia FAQ's response to a question concerning its newswire, which goes as follows: "the Indymedia newswire encourages people to become the media by posting their own articles, analysis, and information to the site," and "Indymedia relies on the people who post . . . to present their information in a thorough, honest, accurate manner" ("Indymedia's").

Thus, it is the openness and heterogeneity of the system combined with the critical analysis and responses of many users that provide backing for credibility judgments by users. On its FAQ, Indymedia explicitly recommends that all users "look at all reports [they] read on the Indymedia site with a critical eye" ("Indymedia's"). On both the producing and the receiving end, the responsibility is in the hands of users—to write accurate, relevant articles of interest to Indymedia's users and to review and evaluate what they read there. This emphasis on personal responsibility and critical thinking is emphasized by commentators such as Ballve (2004) and emerges clearly in comments on Indymedia's FAQ and in responses by users to the site's postings.

Some of the comments in response to the Rios story also illustrate the differences between practices of reading and judging the credibility of mainstream news and alternative news. Readers who are seeking out further evidence, balanced reporting, and editorial control to insure that what they are reading is credible will find that Indymedia's procedures do not guarantee the presence of these elements. Indeed, the sorts of principles that serve to back warrants on this site are specific to the site's emphases on social justice, antiglobalization protest, protection of human rights, preservation of free speech, and prolabor interests. They are not field independent; they are decidedly field dependent, and the credibility of site content depends on users' acknowledgment and acceptance of these principles. Some of them, explicitly stated in the comments responding to various site posts, are that

- freedom of speech and protest are fundamental rights that preserve human equality and freedom;

- protest actions should accomplish more than "just making a statement";
- motivations of corporate entities and most governments are suspect;
- political protesters' rights should be protected;
- the rights of indigenous peoples should be protected;
- education should be available to all, regardless of wealth or social standing;
- refugees and immigrants should not be detained against their will;
- everyone has the right to decent living and working conditions;
- Indymedia news should focus on stories that are repressed or not reported elsewhere.

Such principles have a stronger standing for the Indymedia users than many of the more conventional criteria that are said to provide evidence of credibility, such as

- news should report all pertinent facts;
- news should report both sides of the story;
- quality is insured through editorial control;
- stories should be clearly attributed to an identifiable author or news organization;
- language use should be precise and objectively stated.

The values of Indymedia for many of its users, then, lie in its openness in governance, publication practices, use of open source software and collaboration infrastructure, as well as its decentralization and its willingness to publish a wide array of content. These principles align with its purposes and enable the organization, insofar as possible, to avoid incompatibilities between what it says and what it enables its writers and readers to do.

Conclusion

By examining how context-based criteria enable users to make judgments about credibility on the Indymedia site, I have provided a brief case study to illustrate the roles of backing and field dependency. In doing this, I drew upon scholarly commentaries about the site's practices, the site's own statements in its FAQ pages, descriptions of how the site is run, and comments by users that arose when credibility was at issue.

Having considered how Indymedia's credibility arises from its use of field-dependent standards, we are now in a position to return to the question of what mode of constraint other than authorship might be operating when users must judge the credibility of anonymous or coproduced discourses on the Web. When one interrogates what might be involved in such modes of constraint, it is important to keep in mind that constraints are enabling as well as limiting. That is, they operate to produce and channel judgments as well as to set limits on them. Regular users of Indymedia do not look to the objectivity of its content, the credentials of its writers, or the identity of its editors to secure their judgments of its credibility. Instead, their standards grow out of the alternative media field's commitments to free speech, egalitarianism, and protection of human rights and social justice. The backing in this field thus operates to open up alternative possibilities for deliberation and judgment while closing off those that are more conventional. The notion of field dependence can be shown to function very effectively as a mechanism for explaining how epistemological contexts and the evaluation standards that grow out of them play a role in online knowledge production practices.

This chapter's proposal to look to field-dependent standards as the mechanisms that enable users to make credibility judgments seems reasonable when the present state of online credibility is considered. In light of the wide variety of Web site topics and uses, both empirical research (Fogg et al.; Rieh) and common sense indicate that users will not judge all sites according to the same standards. Observations of user behavior have shown that users apply various standards to different categories of sites depending on their priorities. What needs to be considered, then, is how they go about doing this and what factors influence the credibility judgments they make. I have argued that there is a very reasonable way of thinking about these questions, and that is to consider the field in which the site is located and how it influences user thinking and behavior. Users will not judge the credibility of a medical site in the same way as an entertainment site, an alternative media site, or a travel site. The standards they apply will depend on the characteristics of the field in which the site is located. Therefore, in studying how online credibility operates, recourse to a field-independent criterion such as source expertise is ill-advised. Critics studying rhetorical credibility in Web-based environments should, therefore, consider the role of field dependency in judging online credibility's role as a component of the persuasive effects of public discourse.

Interactivity: The Golden Fleece of the Internet

What does it mean to say that interactivity is the "golden fleece" of the internet?[1] The metaphor seems quite apropos, for like the ram's golden fleece sought by Jason and the Argonauts, online interactivity is something that is highly valued, yet there is little consensus about exactly where it is located or what it is. Is it a feature of the media technology, the communication context, or the perception of users? (Kiousis; McMillan)

Media researchers have viewed online interactivity primarily as an attribute of technological functions of the medium, such as hyperlinking, activating media downloads, filling in feedback forms, and playing online games (McMillan; Stromer-Galley). As Jennifer Stromer-Galley observed, this features-based approach emphasizes media effects and assumes that users are interacting with the medium "without ever directly communicating with another person" (118).

On the other hand, researchers in computer-mediated communication have focused on user-to-user interaction such as is found in email, chat rooms, and on discussion boards. Sheizaf Rafaeli's view is that full interactivity occurs only when messages sustain reciprocal exchanges between communicators. That is, a person sends a message; a respondent replies in terms relevant to the topic initiated by the first person, and then that person responds to the response in a relevant way. (This is like a thread in a series of email messages.) This has been labeled "third order dependency" (Kiousis 359) because the third and later messages in the sequence are related to and extend the topic of the original message. This perspective, then, views interactivity as an artifact of message sequencing and reciprocal communication in the context in which communication occurs.

A third perspective on online interactivity views it as an artifact of what users experience and perceive. Researchers holding this view are primarily interested in how users process communicative messages in online environments. The requisite criterion in this framework is that users must actively attend and respond to messages in order for there to be interactivity. John E. Newhagen noted that those who support the idea of interactivity as perceived by users would maintain that "at least one human has to be engaged in information processing in order for interactivity to take place in . . . 'the true now' " (396).

These three approaches to studying interactivity have been so highly contested that Erik P. Bucy has rightly concluded that "after nearly three decades of study and analysis, we scarcely know what interactivity *is*, let alone what it *does*, and have scant insight into the conditions in which interactive processes are likely to be consequential for members of a social system" (373; emphasis in original).

At this point, readers of this chapter might be wondering what interactivity has to do with rhetoric. After all, in communication studies interactivity is most closely aligned with interpersonal communication. Interactivity is nevertheless a significant linchpin in the rhetorical appeal of online messages. Speculating on the reasons for its importance takes us once again back to the rhetorical theory of Kenneth Burke. Burke believed that persuasion generally is aligned with identification between people. Identification is best promoted through division; that is, people come to identify with common interests by separating themselves from opposing groups and interests. He noted that "we are clearly in the region of rhetoric when considering the identifications whereby a specialized activity makes one a participant in some social or economic class [or, one might add, political party]. 'Belonging' in this sense is rhetorical" (*A Rhetoric*, 28).

Burke also says that "the purest rhetorical pattern [is when] speaker and hearer [join] in partisan jokes made at the expense of another" (*A Rhetoric*, 38). As will be seen in the case studies from MoveOn.org and georgewbush.com later in this chapter, online interactivity played a key role in promoting identification by involving users in parodic discourse about the opposing candidate, encouraging them to join together in defeating the opponent, and organizing vigils and rallies in opposition to the opponent's policies. Such actions bring to mind Burke's observation that "we must think of rhetoric not in terms of some one particular address, but as a general *body of identifications* that owe their convincingness much more to trivial repetition and dull daily reinforcement than to exceptional rhetorical skill" (26; emphasis in the original). On this view, attitudes are formed through such actions as signing petitions, donating to a campaign, activating media clips that ridicule the opposing candidate, and actively advocating for one's own candidate.

Many of these actions are mobilized by means of interactivity and promote identification and therefore persuasion. Furthermore, rhetorical uses of style and modes
of expression can be impersonal, or can themselves have an interactive quality about
them. Thus, online interactivity as a means of activating user response and as a
mode of address can influence users and can itself be rhetorical in its effects.

The purpose of this chapter is to show how online interactivity plays a role in
persuasion by bringing users to identify themselves with the speakers' interests.
Before claiming that a dimension of online interactivity is rhetorical, however,
I revisit the controversy surrounding the nature of interactivity and the forms of its
use. After identifying the senses in which interactivity plays a role in persuasion,
I provide two case studies of opposing political campaigns in the 2004 presidential
election to illustrate and compare their respective uses of Web-based interactivity
as used in one-to-many online communication.

Interactivity: Sorting It Out

One focus of research on interactivity has been its use in political campaign Web
sites. This is because the Web's role in campaigning has grown in importance and
extent since the first online campaign sites appeared in 1994 (Foot and Schneider).
In each election cycle since then, the public's reliance on internet-based political
information has increased. For example, the percentage of citizens citing the internet as one of their main sources of campaign news has risen from 3 percent in 1996
to 11 percent in 2000 to 21 percent in 2004 (Williams). Furthermore, whereas in
2000, only 33 percent of internet users reported that they went online to get news
or information about the elections, this proportion had increased to 52 percent by
2004 (Rainie, Cornfield, and Horrigan).

The importance of interactivity in online political communication has been
emphasized by a number of researchers (Endres and Warnick; Puopolo; Stromer-
Galley and Foot). In a primer for planners of online campaigns in 2002, the authors
noted that "interactivity is one of the great distinguishing qualities of the internet,"
and they advised their readers that to really engage, campaign sites must "take steps
to leaven [their] Net operation's interactive features with interpersonal opportunities"
(Institute for Politics, Democracy, and the Internet 25). To this end, they encouraged
online campaigns to provide the name of a contact person and an offline method to
reach the campaign, include online polls seeking user opinions and post results, provide a sign up for users to subscribe to email updates, maintain a secure online system
for visitors to make campaign contributions, and provide online opportunities for
supporters to volunteer. The primer concluded by saying that "when a campaign

extends interactive features to the public, it signals a willingness to listen and learn from the people. That's a good image for the candidates to live up to" (25).

For those who are interested in studying how interactivity is incorporated into online campaign sites, however, the problem of how "interactivity" is to be conceived and studied remains troublesome. One reason for the hesitancy to work from an interpersonal, message-exchange framework when discussing interactivity in online campaigns was explained in the work of Jennifer Stromer-Galley ("On-Line Interaction"). Analyzing candidate Web sites in 1996 and 1998, she found that most candidates avoided highly interactive features, which she defined as forms of "human" interaction (e.g., discussion boards and email contact with the candidate or campaign). Instead, these Web sites favored "media" interactive features such as basic hyperlinking and downloadable audio and video content.

To understand these patterns, Stromer-Galley interviewed campaign managers and discovered that there were at least three reasons why human interactive features were not used more frequently and in greater quantity among online campaigners. First, managers said that their staffs did not have the time or resources to respond to all of the electronic messages they might receive. Second, campaign managers were concerned about losing control over the campaign discourse. Opening up the site to a wide diversity of views could result in a situation where the media picked up views and responses that did not represent the candidate's own views. Third, specific views that the candidate or campaign expressed on particular issues might result in the loss of "strategic ambiguity" that could mean losing the support of some voters. The need to wisely control media coverage and use campaign resources, therefore, meant that campaigns at that time generally avoided synchronous campaign-to-user or user-to-user interaction on their sites. In 2002, 90 percent of U.S. House, U.S. Senate, and gubernatorial campaign sites avoided using features such as interactive polls, multimedia presentations, individualization of site content, and live, online events. Much higher percentages of these Web sites opted for simpler forms of interactivity, such as online donation systems (77 percent), volunteer sign up (62 percent), and photo galleries of campaign events (46 percent), among others (cf. Foot, Schneider, and Xenos). Furthermore, use of fully interactive mechanisms such as discussion boards and synchronous online chat would open the campaign site to discrepant points of view and possibly cause the campaign to lose its ability to stay on message. Therefore, interactivity between users within the communication context is quite rare on most political campaign sites.[2]

Those interested in studying the use of interactivity on political campaign sites (and other one-to-many venues), therefore, would be well advised to consider the use of interactive mechanisms other than those using online person-to-person interactivity with third-order dependency. In this regard, Rafaeli's discussion of interactivity is once again helpful. He made a distinction between "noninteractive" and two kinds of "interactive" communication. In noninteractive exchanges, subsequent utterances in an exchange are not relevant to earlier messages in their nature or topic focus. Rafaeli contrasted this with "fully interactive" exchanges—message threads in which all subsequent messages were relevant to both the content and meaning of earlier messages. He also included a third category of "quasi interactive" or "reactive" messages in which a person sends a message to another person, who responds, and later messages refer to or cohere with the one preceding them. Classifying user responses to campaign attempts at interactivity as "reactive" is suitable, since most candidate Web sites often opt for a quasi-interactive, one-to-many style of interactivity. This classification of three gradations of interactivity enables campaign site analysts to differentiate "fully interactive" from "quasi-interactive" or reactive responses to interactive features on political campaign sites.

A second task, however, is to consider a form of on-site interactivity that has only recently been studied by researchers interested in interactivity's role and effects on online political campaigns (Endres and Warnick; Warnick et al.) This has been labeled "text-based interactivity," and lack of attention to it may be due to new media researchers' interest in elements unique to the medium. Text-based interactivity nonetheless has been shown to play a role in users' reactions to political Web sites (Warnick et al.).

Text-based interactivity refers to the presence of various stylistic devices, such as use of first person and active versus passive voice; additional visual cues such as photographs of the candidate or supporters interacting with other people; and additional textual content on the site (Endres and Warnick). Such site elements function as rhetorical features of the site text that communicate a sense of engaging presence to site visitors. Campaign sites are purposefully designed to have a persuasive influence on their audience of users, and the use of expressive style, modes of self-presentation, and attentiveness to content has been shown to enhance users' reception of messages and recall of site content (Warnick et al.).[3] It is for this reason that the Institute for Politics, Democracy, and the internet urged campaigns to "extend a welcome greeting," "be concise," and include testimonials and endorsements from citizens not affiliated with the campaign (10). In addition, Cornfield, Safdar, and Seiger suggested that text should be broken up and interspersed with subheadings, photographs, and icons, exhorting webmasters to "add names, faces, quotes, endorsements" because these are elements that make the Web site text lively and engaging (26).

Forms of expression on sites high in text-based interactivity will be dialogic as opposed to monologic. This distinction is drawn from Mikhail Bakhtin's theory of dialogic communication in which he differentiated discourse in which centripetal (centralizing) forces predominate from speech that is multivoiced (decentered). His studies of speech as represented in the novel led him to believe that all contextualized speech was heteroglossic. That is, all situated expressions in their "being said" take their words, styles, and forms of expression from the preceding talk of others. This multivoiced style is to be contrasted with monologic expression which has the quality of "it is said that" and is largely depersonalized. Utterances such as encyclicals, military commands, and official reports are highly conventionalized, depersonalized, and constrained in this way. Markers of such speech include use of the passive voice, denotative words, and third-person address. The passage below, for example, is taken from the 2002 congressional campaign site for Buddy Darden (Georgia, Eleventh District).

> Public service has marked the life of George (Buddy) Darden since early adulthood. Over the years, he has served as a state district attorney, state legislator and—from 1983 to 1994—as member of the United States House of Representatives. Buddy represented Georgia's Seventh Congressional District for six terms. In the House, he served first as a member of the Armed Services Committee and the Interior Committee, then moved to the appropriations Committee and its Defense and Treasure subcommittees. ("About Buddy")

This passage is expressed from a neutral, third-party perspective, is comprised of a set of facts, and is confined to listing the candidate's offices and committee service. In contrast, the following passage from Stan Matsunaka's 2002 congressional campaign site is highly heteroglossic, manifesting a diversity of speech styles, immediate first-person address, use of an illustrative photograph accompanied by an explanatory label, and use of others' speech to express the central idea:

> I am proud to say that I was born and raised in Colorado's Fourth Congressional District, and it continues to be my home today. I was born to Harry and Mary Matsunaka on November 12, 1953 in Akron, Colorado, and grew up in Fort Morgan—another small town on Colorado's eastern plains. . . . One of the formative experiences of my youth was talking with my Dad, who is a decorated veteran of the famed 442nd Regimental Combat Team, the most highly decorated American fighting unit for its size in American history. [Next to this statement is a photograph of Matsunaka's Dad, walking alongside his son and holding a yellow "Matsunaka for Congress" sign.] Their motto was "Go for Broke" a slang term that means to give everything you have to your cause. . . . I have been urged by my family and friends here in Colorado to "go for broke" and . . . I accept the challenge. ("Stan the Man")

In this passage, we have heard the voice of the candidate ("I was born and raised . . .), his campaign organization ("Matsunaka for Congress"), his Dad who holds up the sign with this campaign motto, the motto of his Dad's combat unit ("Go for Broke"), and the urging of his family and friends to "Go for Broke." Through its use of narrative, identification with place and family, and vivid expression, this text conveys a sense of immediacy and qualifies as a form of text-based quasi-interactivity.

In the remainder of this chapter, "interactivity" will be viewed as communication that includes some form of reciprocal message exchange involving mediation and occurring between a group (the campaign) and users, between users and the site text, or between users and other users. It will emphasize the contingent transmission of messages back and forth as well as text-based interactivity as I have just explained it. My emphasis will be on the communication setting of the campaign, the forms of expression in the site text, and the ways in which the site text procedurally enables interaction. I also discuss interactivity's rhetorical effects in political campaign sites on a societal level, taking into account that its effects on groups of individual users have been studied elsewhere (Coyle and Thorson; McMillan; Puopolo; Warnick et al.). My emphasis will be on how text-based interactivity might appeal to site users and on the extent to which subsequent user action vis-à-vis the campaign is made possible online or offline by features of the site text.

A Useful Taxonomy of Interactivity

One of the prominent researchers in online interactivity, Sally J. McMillan has traced the history and emergence of the interactivity concept as it applies to new media. She developed a typology of three forms of interactivity that will serve well in this chapter's case studies of interactivity in two prominent political sites in the 2004 campaign. McMillan identified three forms of interactivity in internet environments—user-to-system, user-to-user, and user-to-document.

User-to-system interactivity is similar to the emphasis on the media technology focus described at the beginning of this chapter. McMillan described it as "computer-controlled interaction [that] assumes that the computer will 'present' information to learners who will respond to that information" (174). In such situations, the user activates a technical capacity of the system, and the system responds. User-to-system interactivity includes clicking on hyperlinks, customizing site features (such as font size and image display), and some gaming operations. Since this chapter emphasizes forms of interactivity insofar as they function as communication rather than as technologically enabled, user-to-system interactivity will play a minor role here.

User-to-user interactivity is communication that occurs between users and is aligned with the computer-mediated communication orientation illustrated in Rafaeli's treatment of full interactivity. Examples on political sites include online town halls, internet chat, blogs with user comments, and moderated discussions. As I have noted, there are good reasons why campaigns are advised to use this form of activity sparingly, and it was rarely found in the two case studies in this chapter.

There is another form of "user-to-user interactivity" (more appropriately labeled Web site author[s] to users, or vice versa). In this case, user-to-user interactivity is somewhat of a misnomer because it brings to mind an image of site visitors *interacting with each other*. In the sites discussed in this chapter, the candidates and their campaign staffs functioned as the points of origin for Web site content and thereby played an authorial role. Campaign-to-user or user-to-campaign interactivity includes campaigns' efforts to reach and communicate with their users. Thinking of it in this way expands the user-to-user category to include many other features, such as email between campaign staff and potential voters as well as online invitations to contribute or volunteer. Campaign-to-user interactivity, therefore, includes many-to-one and one-to-many communication where the level of receiver control is relatively low but where there can be reciprocity in the form of user response to the initiating message. In a 2005 study of interactivity on campaign sites (Warnick et al.), campaign-to-user interactivity included a "contact us" email link, information about the location of the campaign headquarters, a Web-based registration appeal, an events notice, and a Web-based contribution feature. Most of these features are initiated by the campaign to which the user responds, and in some cases the campaign acknowledges the user's response.

User-to-documents interactivity in a new media context occurs when recipients of the message contribute texts and information that change the content of the site text. In this form of interactivity, users become active cocreators of messages when they customize site content, vote in online polls, submit questions to be answered on the site, or post messages and photos that become part of the Web site text. In this sort of exchange, the Web site invites users to submit content; users send in materials; and then those materials are posted to the site for others to read.

The case study analyses in the chapter examine the uses of interactivity on political sites in a communication context and therefore focus on campaign-to-user, user-to-document, and text-based interactivity. As noted earlier, the third of these forms of interactivity focuses on features of expression in the Web site text such as style, forms of expression, and use of visual images intended to enliven expression and engage user interests and response.

The 2004 Election and Online Campaigning

The 2004 campaigns on the Web were noticeably unlike those that preceded them. The campaign season commenced with Howard Dean's dramatic run for the Democratic Party's presidential nomination. Dean's campaign raised over fifteen million dollars online through the fall of 2003, and, by using meetup.com, had registered nearly 140,000 campaign supporters (Wiese and Gronbeck). Blogs developed a stronger presence than they had had in prior elections, and they were used on campaign Web sites to provide commentary (albeit mostly from campaign officials themselves). There was also a sea change in campaign fund-raising practices due to the Bipartisan Campaign Reform Act of 2002 which I discuss later in this chapter. In short, the increased interest in offline mobilization, intensified the use of campaign Web sites to mobilize supporters, and greater emphasis on online resources for fund-raising meant that the Web played a larger role in political campaigning than it had in prior elections. As Wiese and Gronbeck have observed, "the 2004 campaign will be known as the election in which presidential campaigns caught up with Web development and design strategies to produce a complex form of online political communication" (218).

The remainder of this chapter focuses on the rhetorical uses of interactivity as used in two prominent 2004 political campaign sites—moveon.org and georgewbush.com. As its Web site reports, the MoveOn family of organizations was set up to bring grassroots Americans into the political process. It is comprised of two segments. MoveOn.org Civic Action is a 501(c) (4) nonprofit organization that emphasizes education and advocacy on important national issues. MoveOn.org Political Action is a federal PAC that seeks to mobilize citizens to influence Congressional decision making and to elect candidates whose issue stances reflect the progressive, liberal orientation of MoveOn ("What Is MoveOn?"). Georgewbush.com was the main Web site of President Bush's second campaign for president. Its focus was primarily on mobilizing supporters to donate to the campaign, host events, and spread their support messages to local talk radio and newspapers through viral transmission. The sections that follow describe the various uses of interactivity in 2003–2004 on these two sites, examine their rhetorical effectiveness, and highlight their strengths and weaknesses in this area.

Online Interactivity on MoveOn.org

MoveOn.org was founded by Wes Boyd and Jean Blades in 1998 at the time of the Monica Lewinsky scandal to circulate an online petition calling on Congress to censure President Clinton and move on to deal with the issues facing the nation. Some time later, after September 11, 2001, MoveOn initiated a peace campaign calling for a restrained, multilateral response to the attacks that was signed by over half a million people. As of October 2004, the site claimed to have 2.8 million members (Cha). In December 2005, this number had grown to 3.3 million members ("About the MoveOn"). The site's primary strategies have been to circulate online petitions calling for governmental action, coordinate email mailings and Web casts, and organize offline events in support of its causes. In June 2003, MoveOn organized an online primary among Democratic candidates; Dean, Kerry, and Kucinich were the top vote getters in online voting at that time. In July 2004, it organized house parties nationwide in support of Michael Moore's film *Fahrenheit 9/11*, and 55,000 participants took part at 4,600 locations. In September 2004, at the height of the presidential campaign, the site invited users to host a watch party in support of "Vote for Change," and over 700 users responded by organizing parties that users could locate through a search feature on the site ("Attend"). The site steadfastly opposed George W. Bush in the 2004 election.

Since the 2004 election, MoveOn has continued to have substantial success in mobilizing its users on political issues, and on-site interactivity has played a significant role in that success. For example, the organization has created dozens of print and video political advertisements and supported their production. MoveOn then posts the ads on its site and invites users to contribute to a fund so that the ad can be aired. In December 2005, MoveOn used this strategy to raise over 350,000 dollars to finance the showing of a televised advertisement about the Bush administration's Iraq policy. The organization also has organized events and then invited users to host them. For example, in 2005, it urged users to buy or rent a movie about Wal-Mart's exploitive employment and marketing practices and then host a viewing party to see the film ("Recent Success Stories").

Such activities illustrate one of MoveOn's major strengths, and that is to use online mechanisms, such as movie viewings and vigils, to mobilize offline action and to enable users to come together and take action in the material world. It initiates events by providing online infrastructures that furnish information about event locations and planning. The site then encourages participants to submit photos taken at these events and posts them on the site for other users to see. As I explain later, this process intensifies the feeling of on-site interactivity experienced by users.

MoveOn also encourages users to tell their own personal stories of events such as peace vigils and candlelight protests. For example, from materials sent in by participants in "vigils to honor 2,000 killed in Iraq" in late October 2005, MoveOn assembled a page of photos and personal stories to commemorate the vigils. This page featured many photos of individuals and participating groups, and some of the experiences described by these participants were quite moving. For example, a user named David from Urbana, Illinois, described the responses of families of servicemen and women who attended the vigils:

> We were so fortunate to have a Gold Star family with us, and the mother spoke of her son and how proud she was of what he had accomplished in his life. She spoke of his honor and heroism—and how she felt compassion for the other 1999 families who had had the same experience they had had. She expressed the wish that all of the fallen be honored by remembering and supporting and honoring those in the military who continue to serve—remembering the risks they are taking to serve the rest of us. In addition, it was possible for the military families to explain about the isolation they feel and the public's apparent apathy over the war because it doesn't touch their lives on a day-to-day basis, and also gave them the chance to tell of instances of great character, courage, and compassion that they've witnessed in our servicemen and women. ("Vigils")

David situated his account of the vigil by reproducing the speech of members of military families who were there and who talked about their losses and their feelings for others in the same situation. By speaking in "another's words," David's statement brought home to his audience of site users what it was like to be present at the vigil. This and 83 other comments from participants in the more than 1,300 candlelight vigils were posted to this "shared experiences" link to be read by other users. Content such as this is strong in text-based interactivity and enhances a sense of presence and participation in MoveOn's events.

The process of organizing and coordinating vigils incorporates a variety of the forms of interactivity. The site initiates the process by calling for vigils to be held (site-to-user interactivity), and then users respond by clicking on a link and finding out how to participate (user-to-site interactivity). The sequence enables users to join together to represent their views in public (offline user-to-user interactivity). After the event, users send in their photos and experiences (user-to-site interactivity), and then other users can see what has occurred and be influenced by it (text-based interactivity). The photographic and verbal forms of personal expression have the potential to make a lasting impression on the MoveOn user community and to keep them involved in the causes to which MoveOn is dedicated.

On a more light-hearted note, MoveOn mounted one of the more widely publicized efforts at political humor during the 2004 presidential campaign. The site authors explained their motivation for organizing such a contest as follows:

> For the last three years . . . the President has managed to hide behind a carefully constructed "compassionate" image. As the 2004 election nears, it's crucial that voters understand what President Bush's policies really mean for our country.

> That's why we decided to launch *Bush in 30 Seconds*, an ad contest that's intended to bring new talent and new messages into the world of mainstream political advertising. We're looking for the ad that best explains what this President and his policies are really about—in only 30 seconds. ("Why We Did This")

Users submitted hundreds of 30-second spots in digital video and Flash to MoveOn. In response to this call and after minimal vetting to weed out clearly inappropriate or distasteful submissions, MoveOn's organizers posted 1,017 of the amateur commercials in December 2003 and invited Web site users to select their favorites (Eaton). Within the first two days of voting, more than 50,000 people had made 700,000 rankings. A panel of celebrity judges then rank ordered the 15 most highly ranked ads (Bancroft).

These amateur videos did a rhetorically sophisticated job of lampooning Bush's foreign, domestic, and economic policies. The top winner, "Child's Pay," showed small children hauling garbage and mopping hallways and ended with the tag line "guess who's going to pay off President Bush's $1 trillion deficit?" (Bancroft). Another ad showed a man listening to news about 87 billion dollars to provide health care and education in Iraq; the man then raises his eyebrows and speculates, "Maybe we can get him to invade here."

"Bush in 30 Seconds" is an example of user-to-document interactivity by means of its use of user-created videos and invitations to vote on the videos. This sequence extended the online event to include user-to-site and site-to-users interactivity over a period of weeks. Morris Reid, a prominent communications consultant for political campaigns, emphasized the importance of MoveOn's continued interaction with its users:

> Inviting people not only to make their own ads but to help pick the winner was a brilliant stroke. . . . Everyone wants to come back the next day and see who won. Americans love that. And [the site is] really smart about staying with their strength. . . . They know their members have to feel they're continually being brought into the process. (Cited in Bancroft)

The level of activity and kinds of interactivity to be found on the site vary according to what is going on in the larger political context. Consulting site versions posted in October and June 2004 when the presidential and congressional campaigns were under way revealed a wide variety of forms of interactivity on the site.[4] These not only included the ubiquitous, standard forms of site-to-user interactivity (e.g., "Become a volunteer," "Contribute," "Sign up for action updates," "Sign petition," etc.), but also included calls for user response specific to the campaign season. For example, users were encouraged to vote, to vote early, and to report problems with the voting process. In part, MoveOn was using its site as a mechanism of surveillance for voter abuse or fraud. The intensity and frequency of MoveOn's interactivity with its users diminished somewhat in the months following the election. One year after the election in December 2005, the site showed a lower level of interactivity and fewer interactive links. The site's forum pages where users posted their opinions on MoveOn's future goals showed lower activity, and calls for future action on specific issues were less frequent. Thus, site activity was reduced to standard calls for user response, such as "Contribute," "Sign up," and "Report a problem with the website."[5]

There are a number of improvements that MoveOn could make to enhance interactivity on its site. For example, as of mid-2006, there was no user-to-user interactivity in MoveOn's forum section of the same kind as I described in the "comments" section of Indymedia in Chapter 3 or in the highly interactive blog on the Dean for America site. Instead, the forum posted individual comments from users that the user community then ranked and prioritized, but this is not a mechanism that enables users to respond to other users. Furthermore, it is unclear what the site does with the petitions that it asks users to sign. The users sign off, send in their names, but the site does not inform them about the petition's specific destination. It is also unclear where the videos and other media files whose "airing" is supported by user contributions are actually played.

Supplying such information about what is accomplished with the content that user contributions have supported would lead to a more interactive relationship with them. In addition, MoveOn should really provide contact addresses beyond Web forms and generic email addresses. Furthermore, telling users that site authors "get a lot [of] feedback" and not to "expect a personal reply" does not enhance a sense of campaign-to-user interactivity ("Other Question"). Finally, the site lacks a search engine or a site map. Addition of some of these components might go far in persuading some users that site authors care about them and want to make site use easier.

Online Interactivity on Georgewbush.com

As I have noted, other candidate sites in the 2004 election followed Dean for America's model by offering a variety of interactive features that were being used for the first time in presidential campaign sites. These included widespread use of a Web interface to organize offline events where supporters could come together (called "peer-to-peer event planning"), as well as extensive on-site efforts to gather contributions and use volunteers in small-donor fund-raising (Samuel). In April 2004, georgewbush.com created a party hosting system to organize over 5,000 "Parties for the President" and the site continued to use this event-planning mechanism to promote more parties later in the election season (Malone, "Election 2004"). Johnkerry.com did the same, initiating a relationship with Meet Up that led to over 30,000 parties planned through the internet in October 2004.

Surveys of participants in these events during the campaign season indicated that, with each event, "supporters become more committed, more likely to volunteer in the campaign, [and] more . . . likely to give money" (Samuel). Enabling offline user-to-user interactivity in this way was significant in making it possible for the campaigns to garner contributions from small donors throughout the country. Small donor contributions were particularly vital during the primaries because both Bush and Kerry had turned down government matching funds so as to free themselves from federally mandated spending limits.

The McCain-Feingold campaign finance legislation that went into effect prior to the 2004 election was intended to block six-figure contributions to campaigns from individual, corporate, and labor interests. It was also intended to halt contributions to regulated entities at the same time that legislation of direct interest to them was being considered. The hope was that the new reliance on smaller donations would lead to more emphasis on grassroots financing (Potter).

In the end, campaign finance reform did play a role in an election that involved spending for presidential and congressional races of more than four billion dollars, roughly thirty percent more than four years before. Contributions shifted from very large corporate and union soft money to funds garnered over the internet and through 527 groups. As Julia Malone observed, "hundreds of thousands of Americans, including many who had never given to a political cause, discovered the ease of clicking the 'donate' button on their computer screens. In an unprecedented outpouring they made modest contributions of $25, $75, or $100 to the independent political groups, the presidential candidates, and the national parties." For example, John Kerry's Web site had raised $82 million online by October 2004, whereas George W. Bush's site had raised only $13 million. That the two sites were being used for different purposes, however, can be seen in the fact that by the same date, the Bush site had recruited 1.2 million volunteers, whereas the Kerry site had recruited only 750,000 volunteers (Samuel).

The focus of the remainder of this chapter will be on George W. Bush's 2004 campaign Web site.[6] This site offers a contrast to Move On in its political organization, strategy, and coordination of online and offline activity, at the same time reflecting similar shifts in how the Web was being used during the 2004 election cycle. As I show later, the Bush site was less successful than that of Kerry's in influencing voters' decisions. Considering how georgewbush.com used interactivity on its site may lead to some tentative conclusions as to why users responded to it as they did.

Two probable weaknesses in the Bush site were the absence of text-based interactivity and the design of the site itself. In early January 2004, a number of Web designers were asked to evaluate the official sites of the nine Democratic contenders for the presidential nomination as well as the George W. Bush site.[7] Their assessment of the Bush site was that it was too busy and cluttered, offering three navigation formats that made it difficult for users to find the information they desired. One analyst also observed that the candidate was "presented as being removed from the site, having no direct connection," that he seemed "remote and unapproachable" (McCoy). The analyst contrasted the Bush site with Richard Gephardt's site, which was filled with quotations from the candidate, informal photos with his wife, and a column by his daughter, making Gephardt "appear friendly and approachable" (McCoy). As I noted earlier in this chapter, use of first-person address, illustrative photographs, and personalized speech are markers of text-based interactivity, a component of the rhetorical appeal of the site text in which georgewbush.com appeared to be lacking.

A close examination of the forms of interactivity used in the Bush/Cheney Web campaign reveals some notable similarities to strategies used by MoveOn, as well as some noticeable differences. The Bush campaign had an almost limitless budget to support its campaign activities, over $367 million dollars ("2004 Election"). The site, therefore, was in a position to offer an impressive array of features to encourage user retention, involvement, and mobilization. Regarding interactivity specifically, these included a blog (archived by state, topic, or month), a chat center, an action center, a mechanism for organizing users into specific constituencies, and a feature whereby supporters could organize a "party for the President" as well as other events to bring them together. Opportunities for users to directly address each other or the site audience online were comparatively infrequent, however.

Like MoveOn, georgewbush.com posted many of the Bush/Cheney advertisements on its site. Many of them were anti-Kerry ads, emphasizing Kerry's presumably contradictory statements about foreign and domestic policy (the Kerry "flip-flops"). Unlike MoveOn, many of these advertisements were developed initially for television and then were posted to the site as video clips. Since the campaign had the financial resources to pay for the advertisements' development, production, and airing, the site did not need to solicit contributions from users to support their use. There was no opportunity for user involvement or interactivity in relation to the ad campaign other than an opportunity to view the ads.

Like MoveOn, the Bush/Cheney campaign site called for supportive users to organize events that would bring Bush supporters together, and they also encouraged party organizers to contribute quotations and photos from the parties to the site to be posted for other users to see. Likewise, just after the third presidential debate in Arizona in mid-October, the site called for users to organize a Neighborhood Walk the Vote event to demonstrate support as the election neared ("Bush/Cheney'04"). These events encouraged site-to-user, user-to-site audience, and user-to-document interactivity in much the same ways as the vigils organized by MoveOn. That is, the Bush/Cheney site provided the impetus, the means (through site-based information and instructions), and the post-event publicity to make the most of the event. Furthermore, these efforts provided an example of how online discourse can instigate offline activity. As in the case of MoveOn, the same can be said of the site-to-user efforts to get visitors to vote (through a "register to vote" link), vote, and vote early.

Another effective link between site-based public discourse and offline mobilization was the Bush/Cheney's online "Action Center." This link provided separate sub-links urging users to "Call talk radio," "email friends," "Write newspaper letters to the editor," and "Create customized brochures and posters" ("Action Center"). Each of these links provided specific suggestions and advice for creating one's statement or initiating an action that would disseminate the campaign's messages into local media environments. Unlike MoveOn, which worded its users' petitions for them, the Bush Action Center links encouraged users to formulate their own messages. It thus required greater user involvement than simply signing off on a prewritten petition.

On the other hand, the campaign used seemingly interactive features such as its chat center and blog to control rather than to open up online discourse about the campaign. For example, the "Chat Center" featured appearances at pre-announced times by spokespersons and experts affiliated with the campaign such as Liz Cheney, Jenna and Barbara Bush, Bush Campaign Manager Ken Mehlman, Doro Bush Koch, Campaign Press Secretary Scott Stanzel, and Karen Hughes. The blog feature then would repost responses from these people's statements on its blog, along with many supportive comments from other bloggers or newspapers about the campaign. Aside from questions submitted for the chat events by users, there generally were very few user-contributed comments on the chat and blog links.[8] Instead, much of the supposed "site to user/user to site" interactivity was limited to what the campaign wanted to convey to its user community.

Two exceptions to this pattern occurred when the campaign asked users to explain why they supported President Bush, and also when they were asked about their reactions to his performance in the third debate. On a link that included the supporters' photographs and statements, users provided some heartfelt and specific reasons for their support of Bush. For example, Todd Gobermille of Parkland, Florida, wrote,

I support President Bush because, as working parents of two young girls, we must not only be great role models for our kids, but we must also save for their futures. President Bush's Child Credits have allowed us to put an extra $800.00 to their college funds and he is a President who our girls can look up to and say, "someday, I want to be just like him!" Thanks for being a great leader and role model during the most difficult time in our nations history President Bush! ("I Support")

In a statement such as this, written in the first person and providing a concrete example of how a family benefited from a Bush policy, the high level of text-based interactivity buttresses the immediacy and personalization of expression on the blog.

Joshua Fishman of New York, NY, wrote, "I support the President because he has been steadfast in the face of tremendous challenges facing America and the world. He is a straight shooter and a man of his word." Reactions to the debate included a statement from Christina Corieri of Arizona Students for Bush along with an eyewitness account of events surrounding the debate:

Despite the bus loads of professional protesters and the vulgarity the other side brought in, the Students for Bush never faltered. We know that the behavior of the liberal protesters will only turn off independent voters. We have no doubt that Arizona will remain in the Bush column and have faith that after his outstanding performance last night, that other swing states will ultimately turn red as well, but with only 19 days left until the election, we still need to work hard. ("Official Campaign")

The statements of Gobermille, Fishman, and Corieri fall into the category of user-to-document interactivity in which users have the opportunity to directly address the site audience. They are notable because such presumably spontaneous expression of this kind rarely occurs on campaign sites due to campaigns' desire to control the discourse that visitors see on their sites. Of course, these statements were carefully selected from all those that were submitted, but they did allow some users to express their views in a public forum.

Despite its fairly strict constraints on user-contributed expression, interactivity on the georgewbush.com site had some strengths as compared with interactivity on MoveOn. For example, the "contact us" link on the Bush site listed a mail address, phone, and fax numbers in addition to an email address, thus offering users a number of ways to contact the campaign. Georgewbush.com also provided users with a complete site map, thus making immediate access to its contents available in a form that did not exist on the MoveOn site. The Bush/Cheney site's affordances enabling offline user action, including its action center, events calendar, and mechanisms to host and organize events were of unusually good quality and required concerted infrastructure development and maintenance.

The absence of a forum section and the minimal opportunities for users to contribute content, however, made the Bush/Cheney site a typical campaign site in its reluctance to cede any substantial control of issue-based online discussion to outside or divergent voices. Instead, campaign-to-user interactivity dominated the site. In June 2004, there were 15 nonredundant instances of site-to-user interactivity, 2 of user-to-site audience interactivity, and 4 of user-to-document interactivity on the site's home page. This emphasis on site-to-user interactivity intensified as the election neared. In October 2004, there were 19 nonredundant instances of site-to-user interactivity, none of user-to-site audience, and only 3 of user-to-document interactivity on the site's home page. There were no instances of user-to-user interactivity in the top level links on either the June or October 2004 versions of the georgewbush.com site.[9]

The Future of Online Campaigning?

Following Howard Dean's online/offline presidential campaign in 2003 and 2004, the major party candidates' turn to blogs, use of user-contributed content, and efforts to organize offline events was quite noticeable. There continues to be speculation about the shape of future online campaigns. For example, increased use of interactivity is almost inevitable. A 2006 article in the *New York Times* predicted more extensive use of email, blogs, and text messaging to raise money. The article also emphasized possible use of podcasting, viral attack videos, online social networks, and the practice of beaming videos directly to cell phones (Nagourney). In considering these technological developments, political consultant Jerome Armstrong observed that "the holy grail that everybody is looking for right now is how you can use the internet for persuasion" (cited in Nagourney).

Other offline developments may also fuel intense interest in the persuasive dimensions of political and candidate sites in the future. Shifts in campaign financing practices that encourage campaigns to turn more to small donors for funding may intensify use of the Web as a site for soliciting contributions to support the campaign. Furthermore, the extent and prevalence of mail-in and absentee balloting are increasing. At-home voters have the opportunity to seek out information about candidates as they are filling out their ballots. They have the option of going to political information sites that include embedded links to individual candidate Web sites. It is known that undecided voters making last-minute choices look primarily for candidate stands on issues and biographical information about the candidate. There is often no source more useful for quickly gathering such information than the candidate's Web site. Campaigns that put some effort into enhancing interactivity on their sites can increase the likelihood that their visitors will remain on the site longer and better retain the information they find there, since interactivity produces these effects (Warnick et al., "Effects").

Conclusion

Interactivity on the two campaign sites studied in this chapter appeared to function rhetorically to advance the agendas of their respective political interests in a number of ways. In both cases, interactive aspects of site content functioned to join people together in common cause against an opposing candidate and thus promote identification (and, Kenneth Burke might add, persuasion).

MoveOn's "Bush in 30 Seconds" event was especially oriented toward depicting Bush in a negative frame and enlisting users to do it. The user-created-and-contributed anti-Bush advertisements were intended to shift the frame for Bush's presidency from the Iraq War and national security to his domestic policies. The most highly ranked ads emphasized his administration's record of increasing deficits, diminishing social services, and providing tax cuts for the wealthiest Americans while defunding entitlement programs. As I have noted earlier in this chapter, MoveOn's decision to recruit users to engage in a highly interactive process of conceiving, producing, posting, and selecting the best ads insured continued repeat visits to the site and garnered a good deal of media coverage.

Instead of using humor against an opponent as the MoveOn site did, the Bush/Cheney campaign used its "Kerry Media Center," a venue where Republican campaign planners made every effort to call into question the words and actions of their opponent. In mid-October 2004, the "Kerry Media Center" page invited users to view its anti-Kerry television ads, read about numerous instances in which Kerry voted for higher taxes, browse negative assessments of Kerry's performance in the third Presidential debate, and consider ways in which Kerry's claims about Bush policies were misleading. In addition, there was consistent coverage of the "Kerry Flip Flop of the Day," an effort to portray the Democratic candidate as indecisive and confused ("Kerry Media Center"). While it is difficult to determine how effective this Web-based strategy was, the Republicans' one-to-many, noninteractive approach to casting doubt on Kerry's record seemed designed to overwhelm users with information rather than to involve them in any form of user-to-document interactivity as did other portions of the Bush/Cheney site. On the other hand, this trend toward negative construal of the opposing candidate aligned with general trends in both campaigns. For example, in a study of news releases posted to candidate Web sites in the 2004 election, Souley and Wicks found that "80% of the news releases posted by George Bush and John Kerry on their campaign Web site contained an attack against the opposition" (535).

A second area in which political sites such as MoveOn and georgewbush.com have fruitfully used interactivity for rhetorical effect is in furnishing forwardable materials such as video clips, campaign paraphernalia, and statements by the candidate and other supporters that can be disseminated through a process labeled "fan-out" to other supporters. A study of the 2004 election estimated that over 17 million Americans sent emails and other materials about the campaign to groups of family members and friends (Rainie, Cornfield, and Horrigan). Foot and Schneider emphasized the importance of such activity when they noted that "the aim of mobilizing is to expand the campaign exponentially by motivating and equipping supporters to promote the candidate within supporters' social networks and spheres of influence, such as neighborhoods, workplaces, organizational settings, and local media markets" (132). Content posted to the site accompanied by encouragement to share it with others extends site-specific news and information to other online audiences through user-to-user interactivity.

I have already noted a third type of interactivity used for rhetorical purposes, and that is development of offline peer-to-peer networks through event creation such as peace vigils and protests on MoveOn and rallies and parties for the president on georgewbush.com. In 2004, such events frequently brought together groups of people who lived near each other and/or shared common interests, and they provided them with an opportunity to work together offline to support the campaign, thus providing the impetus for face-to-face, user-to-user interactivity and increased allegiance to the interests of the campaign. The significance of this potential for new media to enable formation of constituencies was noted by Manuel Castells when he observed that

> [t]he fact that the audience is not a passive object but an interactive subject opened the way to its differentiation, and to the subsequent transformation of the media from mass communication to segmentation, customization, and individualization. (*The Rise*, 337)

Castells's statement emphasizes the benefits to be had from organizing one's constituents into coalitions by ethnicity (African Americans, Asian Pacific, Hispanics), professions (small businessmen, ranchers, first responders, health professionals), and interest groups (conservative values, seniors, sportsmen, students, veterans, W stands for women) as the georgewbush.com site did. The campaign is then in a position to convey targeted messages to each of these groups that are well designed to appeal to their priorities and interests.

Interactivity on candidate and political sites has other rhetorical effects. By involving users in posting and reading user-contributed content, campaign sites can promote user/candidate identification and thus intensify loyalty to the campaign. High levels of text-based interactivity in the form of direct messages from the candidate, photographs of the candidate with his or her family or interacting with voters, use of events notices, and other means give the impression that the candidate is responsive and accessible. It could be that use of carefully moderated forum sections or discussion boards would further advance an impression of candidate accessibility as well, thus promoting deliberation and discussion of policy topics.

In general, however, we can say that MoveOn and georgewbush.com used elaborate feedback structures on their sites to involve users in establishing what the users saw, read, and heard. To the extent that they succeeded in this process of involvement, they also succeeded rhetorically in deepening the users' sense of affiliation with and commitment to the campaign (Burnett and Marshall; Foot and Schneider). Thus, major advances in user involvement in 2004 were based on offline events planning and increased use of user-contributed content (such as the "Bush in 30 Seconds" videos) on the two sites.

Nevertheless, Stromer-Galley's original observations about the use of interactivity on political campaign sites still hold. She noted in 2000 that "political candidates are using their websites in ways similar to television or radio advertisements—as one way messages," and she concluded that "candidates utilize the media-interactive features to create an appearance of [user-to-user] interaction" (116–117). MoveOn failed to allow user comments in its forum section, and this made deliberation and disagreement about its policy stances infrequent on its site. The site called for advocacy and action through petitions and support for ads it had created, but it sometimes failed to inform contributors about the distribution of the petitions they had signed or the placement of the ads they had financed. Although MoveOn promised that "a real person" would reply promptly to press queries, it then told regular users not to expect any reply to their comments to the site. Thus, most of the communication on MoveOn was one way—from MoveOn to the users—and the site was clearly focused on influencing public opinion rather than enabling public deliberation.

On the Bush campaign Web site, "interactivity" was a shell for meeting other goals—getting favorable press coverage, mobilizing existing supporters, and maintaining focus on its overall campaign message. This lack of commitment to actual interactivity could be seen in the severe limitations on user-contributed content on the site. Quotations in the site blog were posted by the campaign and known supporters. In its chat room, populated by the Bush family and supporters, the questions that were submitted were carefully screened, and answers were selectively clipped and reposted in the site's blog. Polls on the Bush site were used to test users' knowledge of President Bush's existing policy stances rather than to obtain user opinions. On both of these sites, then, a dynamic and potentially two-way medium was made to function much like television, and the focus was on getting out a predetermined message and on obtaining resources for the campaign's use.

By using forms of interactivity that genuinely sought out users' views and then incorporating them into policy development and issues stances, these campaign sites could have made the most of an interactive "pull medium." A pull medium is one in which users actually seek out and construct their texts for consumption rather than being passive recipients (Bucy). Both MoveOn and georgewbush.com could have included user polls that sought out users' viewpoints and either summarized them or posted them to the site. They could have used blogs that were collaboratively authored as were Howard Dean's.

The fact is that neither of these two campaigns aspired to including full interactivity on their sites. The reasons for this absence are likely the ones that Stromer-Galley has described: lack of sufficient resources to handle individual submissions, concerns about losing control of campaign discourse, and loss of the strategic ambiguity needed in appeals to a mass audience. The finding we might take away from this is that the forms of interactivity used in online, Web-based communication may grow out of the specific communication context and the constraints and demands embedded in it.

It is interesting to note, for example, that the Bush campaign derived only 14 million dollars (less than five percent of its total receipts) from its campaign site. Furthermore, more than 50 percent of those who voted for Bush had made up their minds prior to January 2004, and another 20 percent had decided prior to the conventions (Rainie et al.) It is possible, therefore, that campaign strategists were using the site for the primary purposes of mobilizing the already converted, keeping them committed as volunteers and donors, and recruiting them to appeal to other people. If so, the function of the site would suggest an epideictic form of celebratory discourse designed to sustain, retain, and reinforce the solid support already based in the Bush camp. In that light, one can say that the forms of campaign-to-user and user-to-document text on the site probably functioned well to meet campaign needs. An implication of this for rhetorical critics is to begin by considering what rhetorical strategies would best suit a given context and then to examine the forms of interactivity that align with those strategies.

5

Intertextuality and Web-Based Public Discourse

An occasional enticing feature of Google's home page is its holiday logos, designed to embellish the plain "G-o-o-g-l-e" heading and remind the repeat visitor of special events and happenings. In early 2006, their archive of these included the heading with a wig draped over the second "o" and a music scale (Mozart's birthday), a version of the header embellished with shamrocks (St. Patrick's Day), and an initially indecipherable series of colored dots (Louis Braille's birthday) ("More Google"). At the time that these are posted on the Google main page, there is no evident explanation for what they mean. With the exception of obvious holidays such as Christmas and Thanksgiving, the user then is tempted to pause, suspending his or her search while figuring out what the intertextual reference might be. There may be a certain pleasure when the user returns to the site later and already understands the clever reference. Google's use of these intriguing images, which can only be understood when the user recognizes the allusion that gives them meaning, provides an example of the rhetorical workings of intertextuality.

This chapter focuses on the rhetorical dimensions of intertextuality as used on the World Wide Web. It considers how the presence of intertextuality may contribute to a site's appeal as readers participate in the construction of textual meaning. This chapter also focuses on specific intertextual references as a form of intertextuality, although, as I noted in Chapter 2, the term can also apply to broader conceptions of what intertextuality is.

In Chapter 2, I also discussed the fact that many extended or complex argumentative forms (with the exception of the enthymeme) require the reader or listener's sustained attention to continuous text or speech. These forms involve reasoning chains in which prior acceptance of earlier claims enables arguers to build on what has been established earlier. Web reading, however, is discontinuous and fragmented; readers read rapidly and piece together what they read from various sources (Kaplan). They are restless. In such an environment, intertextuality has an appeal. It is modular and does not depend on sequenced text. It offers a wide repertoire of ways to engage attention as readers become complicit in constructing the meanings of the texts they encounter.

As this chapter shows, intertextuality depends on other texts, be they comprised of the cultural matrix of the reader's experience and general knowledge, embedded hyperlinks supporting the intertextual reference, salient current events known to the public, allegorical references, the intratextual environment in which the reference is located, extratextual events the reader might research, or humorous forms such as parody and satire.

The purpose of this chapter is to consider the nature of Web-based intertextuality in relation to its rhetorical function. By focusing on various specific forms of intertextuality and explaining how they function online, this chapter identifies strategies used by Web authors drawing upon intertextuality as a resource. Likewise, it considers the probable roles of Web users as readers when they interpret and are influenced by the texts they encounter. After a brief account of the history and use of the term "intertextuality," the chapter considers the forms it takes and the stages readers go through in textual appropriation. Through use of example texts, this chapter illustrates how different readers can interpret the same intertextually based text in different ways depending on their prior experiences and awareness of pertinent facts and events. I also show that one implication of variable reader uptake of intertextual cross-references is that Web authors should consider their likely audiences' knowledge of and interest in public events when they create multimodal, hypertext-based public discourse.

Intertextuality Then and Now

It should be kept in mind that intertextuality is not a new form of expression, nor is it unique to the internet. While the term "intertextuality" was coined by Julia Kristeva shortly after she arrived in France in the mid-1960s (Allen), its variations—parody, caricature, allusion, appropriation, and the like—have been with us for some time. For example, during the medieval period, readers and listeners viewed the literal meaning of scriptural texts as a way into understanding deeper meanings. They drew on allegorical interpretive methods to comprehend signs embedded in the text that required interpretation (Soffer). In the early modern and modern periods with the rise of the novel, intertextuality took a different form. Novelists sought to create texts with references drawn from other texts as well as out of the reader's social context, and they depended on the reader's familiarity with other literary works. Graham Allen has noted that these authors did not just select other texts for appropriation but also selected plots, aspects of character, and ways of narrating from previous literary texts and the literary tradition.

In contemporary times, electronic media content has become rife with intertextuality. Television programs include allusions to off-air events, prior films, and other aspects of the larger textual context (Fiske). Brian Ott and Cameron Walter have described practices of music production that allow rap and hip-hop artists to include samples of other groups' music, fragments of the larger cultural discourse, and even partial lyrics from other artists in their own productions. Electronic music composers and producers are using new technologies in software and hardware to blur the line between appropriation and composition, and listeners who are aware of the cultural references embedded in the music can read sociopolitical messages between the lines.

One example of this is the genre of music entitled the mashup. Music such as this is made up of small (not always recognizable) loops sampled from existing pieces of music and warped together using revolutionary software like Ableton Live ("Live") and Sony's Acid ("Acid"). Loop-based warping programs such as these enable music of any tempo or pitch to be merged together seamlessly. Music or audio from any source, be it speeches, monolog taken from films, song lyrics or acoustic rhythms, can be sampled and stretched to fit existing musical works, without any apparent evidence of this change. Due to the facility with which DJs can remix vinyl, CD, and .mp3 records coming from varying sources, it is possible for producers to target meaningful intertextual content that is only decipherable to a chosen audience and cultural group. Baruh Lemi has noted that in spite of this potential for creative democracy, "the way the members of the recording industry have responded . . . tends to diminish the flexibility with which individuals use digital music" (67).

Producers who are not careful with their appropriations can face severe penalties from a record industry struggling against mass copyright infringement due to P2P networks and bittorrent seeding. In a recent high-stakes example, illegal samples used by industry icon Sean Combs caused the awarding of 4.2 million dollars in damages to the original owners of the audio material ("Judge"). The fact that this sample is still recognizable is not usually a concern for mashup artists, who are processing them so heavily with effects that their original source is often questionable.

In this way, they can achieve intertextually effective communication without added fears of reprisal.

Another way in which content is being mashed up is in the appropriation of Web content from existing sites in order to make one's own. There are multiple ways of achieving this goal. Some users copy the code of a site and paste it directly into Dreamweaver. Others may use Web spider or crawler programs to capture sites and all links associated with those sites for offline archival storage and manipulation. Web archival programs enable Web sites to be captured as editable text and image files that can be altered, combined, and republished. Screenshots of sites can be used as raw material for new pages, and programs such as Snagit make work such as this seamless and fast ("SnagIT"). The fact that sites can so easily be copied is problematic from an intellectual property perspective, and copyscape enables Web designers to search for copies of their code on the Web, and post plagiarism warning banners on their pages ("About Copyscape"). As the advent of the promise of Web 2.0 emerges, the interactivity heralds even more intertextuality from the appropriation of content from a multiplicity of sources.[1]

A second, politically oriented example of the rhetorical use of intertextuality can be found in audio and video collages of the speeches and remarks of major political figures, such as can be found on DIY (Do It Yourself) media at <http://www.diymedia.net/collage/truth.htm>. Here one can find a few video and hundreds of audio collages of political speech produced by amateur new media buffs. These include, for example, versions of President Bush's verbal gaffes, verbal parodies of his State of the Union speeches, and raps on John Kerry's nonposition on the war in Iraq. Viewers familiar with these political figures and their ways of speaking are most likely to find these versions of parodied speech amusing.

Hyperlinking also enables Web authors to readily link to external information (such as news reports) that supports their parodic content. Parodic allusions that the reader does not quite understand can pique his or her interest and cause the user to follow embedded links so as to obtain further information about the event or situation being parodied. This background information supports the parody and has an intertextual relation to it. Furthermore, the more familiar users become with the situations being parodied, the more likely it is that they will be able to track further intertextual references throughout the site. This keeps the user on the site longer and furthers the potential of the site to keep her involved and thus influenced by the persuasive messages found on the site.

The open, malleable structure of Web site production enables site authors to invite users to submit their own parodic material, as was so well illustrated by MoveOn's "Bush in 30 Seconds" campaign described in Chapter 4. Use of intertextuality is also made easier by the ready availability of audio and video clips of people's actual behavior. Site authors can juxtapose a selected clip with parodic commentary, or even place one clip alongside another clip showing contradictory behavior.[2] Digital alteration of photographs, audio, and video thus has served as a rich resource for the development of online, intertextually based parody.[3]

Another Web-based content platform is Flash. Flash is a popular multimedia Web authoring program that uses vector and raster graphics, a scripting language, and streaming video and audio to create animations. The term "flash" refers to the authoring program, the browser plug-in, or the application files (Kay). Due to the ease with which various objects and displays can be represented, Flash and its companion application, Fireworks, function as killer applications for production and reception of Web-based satire and parody (Kay). Flash's use of a timeline-based approach to defining what happens onscreen makes it possible to produce dynamic content at comparatively low cost. Flash's popularity is also due in part to its small file sizes and its player's capacity to initialize very quickly. Thus, the malleability, modularity, and dexterity of digital production and display technologies have combined to create a rich breeding ground for use of intertextuality in public discourse in the World Wide Web.

Early Conceptions of Intertextuality: Kristeva and Bakhtin

To some extent, Graham Allen is correct to have observed that "intertextuality is one of the most commonly used and misused terms in contemporary critical theory" (2). As I explain later in this chapter, the term is so unstable that nearly every theorist or critic writing about it defines it in a different way. To say that the term is "misused," however, implies that there is a "correct use" against which other definitions can be gauged, and that appears not to be the case. The best hope for gaining an appreciation for the term's many meanings and applications is to consider the various ways that it has been specified and defined since its inception, and that is I what I plan to do.

At the broadest level, one might think of intertextuality as "the fact of one text including various references from another text or texts" (Hitchon and Jura 145). Thus, intertextuality occurs when one text is in some way connected in a work to other texts in the social and textual matrix. There are, however, so many ways in which this can happen that there is little consensus about what constitutes intertextuality. In this section, I plan to begin with the broadest and least limited idea of the term and then proceed to describe versions of intertextuality that are more specific. In light of my purpose in this chapter, I have selected those conceptions that I believe to be the most useful for analyzing the rhetorical workings of intertextuality in Web-based discourse.

As I have noted, the term "intertextuality" first entered the critical lexicon through the work of Julia Kristeva (Allen). Kristeva had immigrated to Paris from Eastern Europe where she had been strongly influenced by the writings of Russian literary scholar Mikhail Bakhtin. Both Bakhtin and Kristeva were centrally concerned with the ways in which discourse was presented and experienced in the novel. Bakhtin viewed the novelistic literary word as an "intersection of textual surfaces" (Kristeva 64) that interwove the speech of the writer, the characters, and the ways of speaking of a given time period artistically into a narrative whole. Kristeva believed that the novelist's writing unfolds and faces in two directions—toward the narration of the work and also toward its textual premises (i.e., speech drawn from other texts and the speech context). Her conception of intertextuality growing out of Bakhtin's work thus was diffuse because, in her view, novelistic works were comprised of many voices and texts and were thus multivocal. Like Bakhtin, Kristeva assumed that "any text is constructed as a mosaic of quotations; any text is an absorption and transformation of [other] texts" (66).

This characterization of intertextuality has been widely misunderstood as displacing the author's intended meaning and giving over control of the text's interpretation entirely to the reader (e.g., Irwin). Bakhtin's intent, however, was to de-center the novelist's use of speech and language. He viewed the novelist as one who represents speech, as a writer who "does not speak in a given language (from which he distances himself to a greater or lesser degree), but [who] speaks, as it were, *through* language, a language that has somehow more or less materialized, become objectivized, that he merely ventriloquates" (Bakhtin 299; emphasis in original). For Kristeva, then, intertextuality was the space in which "in a given text, several utterances, taken from other texts, intersect and neutralize one another" (36).

This idea might best be understood from an example used by Bakhtin from Charles Dickens's novel *Little Dorrit*:

> The conference was held at four or five o'clock in the afternoon, when all the region of Harley Street, Cavendish Square, was resonant of carriage-wheels and double-knocks. It had reached this point when Mr. Merdle came home *from his daily occupation of causing the British name to be more and more respected in all parts of the civilized globe capable of appreciation of wholewide commercial enterprise and gigantic combinations of skill and capital.* For, though nobody knew with the least precision what Mr. Merdle's business was, except that it was to coin money, these were the terms in which everybody defined it on all ceremonious occasions, and

which it was the last new polite reading of the parable of the camel and the needle's eye to accept without inquiry. (Book I, ch. 33; cited in Bakhtin 303; emphasis in the original)

The first sentence of this passage is a colorful but factual description of the conference site in the author's voice. The italicized speech that follows it is a parodic stylization of formal speeches of the day. Spoken in a ceremonious epic tone, it draws from presumptuous expressions then circulating in the social discourse. The author's voice then returns, only to be interrupted again by a wry aside—"these were terms in which everybody defined it on all ceremonious occasions" (303). The passage concludes with a specific intertextual allusion to Matthew 19:24 where Jesus said "it is easier for a camel to go through the eye of a needle than for a rich man to enter the kingdom of God." In this passage, then, the author speaks in his own voice when providing descriptions of narrated contexts and events but then appropriates texts from other sources—officious speech, what "everybody says," the Bible—to represent various points of view. As Bakhtin noted, the novelist's discourse is often thus "another's speech" . . . in "another's language" (313). The adroit use of such speech is what constitutes the artistry of novelistic prose representation. This was Kristeva's view of intertextuality at the time that she coined the term.

It may serve us well to think of intertextuality in this broad sense. The Web (and other venues of internet discourse) are cacophonous environments, sites of Burke's "unending conversation" (*Philosophy*, 110) where many voices blend and clash. They are comprised of a network of textual relations where "meaning becomes something which exists between a text and all the other texts to which it refers" (Allen 1). Here all utterances depend on and draw from other utterances, and every expression is shot through with other competing and conflicting voices. Although speech on the Web may not emulate the artistry of novelistic prose, it nonetheless represents in itself a very unstructured intertextual environment. In the interest of developing a more highly defined typology of forms of intertextuality found in message texts on the Web, however, I now turn to definitions by other, subsequent theorists. My plan is to describe four forms of intertextuality that have been defined in the literature, moving from a general form to other, more specified forms.

Intertextual Variations

Let us begin with a form of intertextuality that establishes a relation between signs and texts on the one hand and the larger cultural text on the other. This form of relation, labeled by Hitchon and Jura as *archetypal allegory*, is not with a supporting "text" per se but instead with an allegory in which characters or events "represent particular qualities or ideas related to morality, religion, or politics" (*Cambridge Dictionary*). These might be characterized as themes or ideas widely recognized in a culture, and Hitchon and Jura observe that while "the average reader usually does not realize where [allegorical] images come from . . . they are . . . perfectly understandable within a particular culture" (147). That is, culturally specific motifs, knowledges, and beliefs provide the intertext that informs the reader's appreciation and understanding of the intertextually supported text.

An appropriate example of this phenomenon can be found in Robert Scholes's book *Protocols of Reading*. Here he describes two images—a painting of two figures and a photograph of a mother giving her daughter a bath. The first of these, entitled *The Education of the Virgin*, appears to be an early seventeenth-century painting in which a girl is reading a book with the aid of a large candle held in her left hand. The book she is reading is being held open by a woman with an expression of calm and patience on her face. Although the book cannot be visibly identified, Scholes speculates that it is the Bible and, based on the painting's title, that the child is a visual representation of the Virgin Mary as a child. The second image, captured by an American photographer in Japan in 1972, shows the female figure in the bath as misshapen, with legs as thin as sticks and arms and hands twisted in impossible shapes. Scholes tells us that this image was part of a photo essay that appeared in *Life* magazine of events in a Japanese fishing village that had been saturated with industrial pollution, causing illness and deformities in the local population. The grown child in the bath is being held by her mother, and Scholes notes that her expression is one of "tenderness and love" (27). He maintains that, in the interpretation of both of these images, the Christian allegory plays a role. For the Western reader and across time and space, the bond between mother and child enhances our understanding of the visual texts such that "those of us brought up in the tradition of Christian art read the picture[s] unconsciously, in terms of this cultural code, which conditions our response" (26).

Allegorical motifs frequently provide an intertextual resource for the representation of ideas and themes in online environments. These may be represented in various ways (e.g., as a quest in search of some prized object, sacrifice of a person or thing for some larger value or purpose, or sudden illumination and discovery because of divine intervention). In any case, we can look to allegory as "a specific method of reading a text, in which characters and narrative or descriptive details are taken by the reader as an elaborate metaphor for something outside the literal story" ("allegory," *Encyclopaedia Britannica*).

A second type of intertextual reference that complements *archetypal allegory* is cross-reference to a *specific film, novel, or other work* that is widely known and recognized by readers and/or media consumers. One example of this type is that of the now-famous 1984 Super Bowl television advertisement for the Apple Macintosh computer that at that time was about to be released (Hitchon and Jura). The spot opens with men, apparently in prison garb, marching in lock step formation into an auditorium where a black and white projection of a threatening, authoritative male figure intones a speech on "Information Purification" and "Unification of Thoughts." Suddenly, a youthful, strong woman in bright red shorts and shoes runs into the auditorium pursued by storm troopers and hurls a sledgehammer at the screen, which then explodes. Then, a calm voice intones that "On January 24th, Apple computer will introduce Macintosh. And you'll see why 1984 won't be like '1984' " (Apple).

As Hitchon and Jura observe, viewers unfamiliar with George Orwell's novel, *1984*, and its account of an oppressive police state and limits on free speech would have been unable to fully appreciate the advertisement's message. They also note that IBM's then-dominant monopoly of the computing market as represented by the persona on the screen was an aspect of the social context supporting the advertisement's message and would have contributed to understanding and uptake of the persuasive aspect of the text's message.

In her study of the ad, Sarah R. Stein noted further possible intertextual cross-references to specific works in this famous advertisement. The opening shot of the marching workers was reminiscent of the 1927 film *Metropolis* that depicted capitalistic oppression and the misery of the working class. The running woman in the ad (the sole colorful figure in a black and white environment) represented "a political figure, one aware of the repressive powers of advanced capitalism and willing to use revolutionary tactics in response" (Stein 187). The sledgehammer-wielding episode and the shattered image of the authoritarian figure on the screen suggested a female David in confrontation with a corporate Goliath. In addition, the bleak landscape of the marching figures and the docile workers in air thick with smog and haze seems visually reminiscent of Ridley Scott's 1982 science fiction film, *Blade Runner*.

A third and commonly recognized form of intertextuality is *parody*, defined as "discursive activity that intentionally copies the style, organization, or other features of a text or situation, making its features more noticeable by way of humorous imitation" ("Parody"). Such a definition implies intentional explicit or implicit juxtaposition of two texts that is contrived by an author who places a popular or known text in relation to another text and that imitates or exaggerates the "original" text.

One example of this can be found in my 2002 book, *Critical Literacy in a Digital Era*, where a text from George W. Bush's 2000 campaign home page is parodied by another parody site that mocks the Bush site's then well-known welcoming statement. The text from the Bush home page is as follows:

> Welcome to georgewbush.com—my virtual campaign headquarters. The most important question I can answer for you is why I am running for President of the United States. I am running for President because our country must be prosperous. But prosperity must have a purpose. The purpose of prosperity is to make sure the American dream touches every willing heart. The purpose of prosperity is to leave no one out—to leave no one behind. I'm running because my political party must match a conservative mind with a compassionate heart. And I'm running to win. (Bush for President, 2000).

The bushlite parody site capitalized on this official home page text by incorporating its style and word choice into a fabricated announcement speech said to have been given in June 1999:

> Prosperity is not a given. That wouldn't be prosperous, nor would it have a purpose. What's the purpose of giving out prosperity to just anyone? Purposeful prosperity—that is prosperity with a purpose—must be earned. To earn it, we need compassionate conservatism. By this I mean conservatism that is also compassionate. (DieTryin.com, 2000)

While capitalizing on the euphemistic quality and circularity of the "original" Bush home page statement, this parody text imports the idea that prosperity must be "earned" through "compassionate conservatism" that is nonetheless not intended for everyone but only for those who "earn" it.

A fourth type of intertextual usage is that which explicitly takes up and plays upon the larger social text. This form of intertextuality frequently takes the form of satire, in which representing a known "thing, fact, or circumstance . . . has the effect of making some person or thing look ridiculous" ("Satire"). *Intertextual satire* is a common form of online expression that draws upon situations widely covered in the news, hyperlinked texts with supporting information, or texts provided elsewhere in a Web site to provide explanatory background for users who do not readily recognize the allusions being made.

For example, shortly after Vice President Dick Cheney accidentally shot Texas attorney Harry Whittington while hunting quail in February 2006, the Web site "Too Stupid to Be President" posted a Flash animation of texts and situations that ridiculed the vice president and his supposed imperviousness to being held accountable for past actions and mistakes. In each case, his "victims" took up the task of apologizing to Cheney for having done anything that might have precipitated their subsequent treatment at the hands of Cheney and his colleagues. The animation began with Harry Whittington, his face and neck full of bird shot, saying "My family and I are . . . deeply sorry for all that Vice President Cheney and his family have had to go through this past week ("Sorry").[4] Whittington was followed by Saddam Hussein, speaking in his underwear from prison, who said, "I am deeply sorry not to have had the weapons of mass destruction, or the plans for them, or the yellowcake [uranium] that Mr. Cheney alleged that I had. I hope that whole unnecessary war thing has not caused him too much embarrassment" ("Sorry"). The next section of the clip shows a hooded figure standing on a box in a poncho and wired to some kind of device while being interrogated. This figure, who "became the indelible symbol of torture at Abu Ghraib prison" [Seattle PI, March 11, 2006], says in a trembling voice:

> Please allow me to extend my apologies to Mr. Cheney (and I'm not just saying this because a CIA subcontractor is raping my children in front of me as we speak.) This is from the heart, bro. If only I had had some scintilla of useful information to offer, I might have somehow vindicated Mr. Cheney's directive to employ torture. ("Sorry")

Finally, there is Valerie Plame, the CIA operative whose identity was leaked in a newspaper column by Robert Novak in 2003. She is sitting in the shadows and appears as a silhouette; at the end of the clip, she is fully illuminated and "exposed" to the light. She says, "I am profoundly sorry that the White House conspiracy to expose my identity, destroy my career, silence other would-be whistle blowers, undermine my husband's credibility, endanger my family, and unleash Iran's nuclear ambitions has in any way reflected badly upon Vice President Cheney" ("Sorry"). The final episode in the Flash animation reveals a set of tombstones, one of which is engraved with the following: "I regret that I have but one life to give to Dick Cheney."

The content of the animation has a number of interesting twists. It ironically contains the apologies of four people for acting in ways that brought down misfortune on them, and it implies that there may be others who will also find themselves in the wrong place at the wrong time and similarly suffer misfortunes because of Cheney's actions. Of the four, only Saddam Hussein is unquestionably culpable; the others are just hapless. The juxtaposition of their oppressed images and their abject apologies adds to the satirical effect. But the most essential force operating in the animation is its use of intertextual cross-reference. Readers who have kept abreast of the news on Cheney's activities would be able to apply news coverage of the quail shooting incident, the Abu Ghraib photographs, and other elements to each of the four parts of the clip, and most of them would find the final play on Nathan Hale's supposed statement, "I only regret that I have but one life to give for my country," doubly ironic, following as it does Plame's apology. (Hale was executed for espionage.)

This description of four forms of intertextuality reveals some of the many ways in which readers themselves play a role in the construction and interpretation of meaning. Understanding and appreciation of the two images in Scholes's description would be enhanced when the images are interpreted allegorically. The Macintosh advertisement was planned so as to be interpreted in light of Orwell's novel, and the parodic speech attributed to "bushlite" could only have been fully appreciated as a parody when read in intertextual relation to the welcome message on the official 2000 Bush campaign site. Only to the extent that visitors to the "Too Stupid to Be President" site knew of the Bush administration's and Cheney's past errors and mistakes could they fully appreciate the satiric commentary proffered by its Flash animations.

Readers, then, play a role in supplying textual readings; they are active participants in the formation of meaning in the texts they encounter. Furthermore, this is a rhetorical process because the more allusions and cross-references the reader gets, the greater is that reader's sense of accomplishment. As Ott and Walter observe,

> Examining such cultural knowledge fosters feelings of superiority and belonging. Since not all [readers] will recognize the allusions, successful identification of parodic references allows readers to mark themselves as . . . literate. . . . The pleasure of recognition is often directly proportional to the difficulty of identifying the allusion. (436)

The next section of this chapter considers how reader reception occurs in intertextual hypertext by explaining how readers encounter sites and make intertextual connections when consuming them.

The Reader and the Intertext

In thinking about the reader's role in interpreting intertextual relations, it is useful to return to Roland Barthes's concepts of the readerly and writerly texts mentioned in Chapter 2. Influenced by Kristeva's account of intertextuality, Barthes emphasized the extent to which the reader plays a role in the formation and comprehension of texts, and he demonstrated that the conventional view of the author as sole origin and generator of meaning was a false one. This author-centered view implied the idea of a readerly text in which the reader is positioned as relatively passive—a figure whose task is to follow the predetermined story line until a truth, presumed to lie behind the narrated events, is unfolded before him or her (Allen). Thus, the "readerly text" is one in which the reader is to be only that—a consumer of the text as designed by the author.

Instead, Barthes viewed the text not as the unique and original creation of an author but rather as "made up of multiple meanings, drawn from many cultures, and entering into multiple relations of dialogue, parody, contestation, [and] there is one place where this multiplicity is focused and that is the reader, not, as was hitherto said, the author" (148). The reader in Barthes's view thus became "that *someone* who holds together in a single field all the traces by which the written text is constituted" (148; emphasis in original). The inveterately intertextual text (as opposed to the unified work) "grows by vital expansion, by 'development' (a word which is significantly ambiguous, at once biological and rhetorical); the metaphor of the Text is that of the *network*; if the Text extends itself, it is as a result of a combinatory systematic" (161; emphasis in original).

This insistence on text as an open, networked system that can be appropriated by readers in various ways introduces some noticeable tensions that must be negotiated in thinking about the workings of intertextuality. If authors and producers design texts so as to be taken up in certain ways by virtue of how they are structured and designed, what role, if any, does the originator of the text have in the interpretations of its meaning? If hypertext productions can be read in any order and connected with other texts in numerous ways, do the authors of Web-based discourse function as authors at all in the traditional sense?

My response to this question aligns with Allen's who, at the end of his long study of theories of intertextuality, concluded that "hypertext makes author, text, and reader into joint participants of a plural, intertextual network of significations and potential significations" (202). It is the case, as I show in this chapter, that both producers and users of Web-based discourse play a role in the constitution of meaning. In most cases of intentionally rhetorical discourse, however, Web authors endeavor to canalize readers' interpretations, but they often do so in such a way that users have a role in the process. Or, as Ott and Walter put it, "some texts . . . deploy intertextuality as a stylistic device in a manner that shapes how audiences experience those texts" (434).

That said, readers nevertheless play a strong role in taking in the texts they encounter, and the question then becomes, how do they do this? It might be helpful here to consider Stuart Hall's tripartite model for how different groups of readers interpret media texts. This includes "dominant reading" in which the reader fully shares the text's code; "negotiated meaning" in which the reader partly shares the text's code but sometimes resists it; and "oppositional reading," where the reader rejects the text's code and brings to bear an alternative frame of reference (Hall). As will be seen in this chapter's case studies, users often encounter sites combining parody, satire, pastiche, general allusions to cultural contexts, and other forms of intertextuality. In interpreting these texts, users must be able to make sense of what they encounter. After all, if the content makes no sense to them, they are likely to break off their reading and viewing and leave the site, so the question of how they make sense of what they see is an important question insofar as the possible rhetorical influence of messages is concerned. The remainder of this section considers how readers come to *comprehend* the multimediated, online texts they encounter and, in particular, what external texts various users might variously bring to bear to understand what they see.

In thinking about how users as readers process the texts they encounter, one might turn to a general account of meaning making in hypermodal environments proposed by Jay L. Lemke. He maintained that there are three semiotic functions that play a role in users' responses to Web-based discourse. The first depends on *presentational* meanings that enable users to construe a state of affairs from what is said, shown, or portrayed on the site. Operationally, I would say that this is the phase of meaning making in which the user, when initially looking at the site, could answer the question "What is this about?"

The second function depends on *orientational* meaning that enables users to orient to the communication situation in terms of point of view. In this phase, a user would be able to answer such questions as "What is being asked of me?" and "How am I being treated or positioned?" Lemke noted that factors such as terms of address as well as the mood or modality of expression work as orientation indicators.[5] The third function is comprised of *organizational* meaning that enables users to determine which signs go together into larger units. Display patterns such as image groupings, link menus, and sites' graphic identity facilitate the users' development of organizational meanings.

In applying this framework to his analysis of two Web sites, Lemke makes it clear that meaning making is an emergent process that commences when the user first encounters the site and then develops his or her understanding based on signs, pathways, forms of expression, and representations encountered on the site. He notes that the three functions are not independent of one another and that we, as users

recognize patterns by parallel processing of information of different kinds from different sources . . . and we refine our perceptions and interpretations as we

notice and integrate new information into prior patterns in ways that depend in
part on our having already constructed those prior, now provisional patterns. (305)

Lemke's account of meaning making here applies especially well to Web-based
discourse that evidences a good deal of intertextuality. Sites of commentary,
resistance, and political parody, as well as entertainment sites involving social criticism may not be immediately understandable to all users. Being able to identify
the relevant intertexts that enable users to understand what is being said is a vital
component in the success of persuasive communication on these sites. Since
different readers possess different levels of textual knowledge on various topics,
they will read and appreciate the same text differently (Ott and Walter). Producers
of such sites also need to negotiate a tension between making their content so
accessible that it lacks originality and uniqueness on one hand, and developing
content with allusions that are so arcane and specialized as not to be understood,
on the other hand.

To illustrate how such considerations can be brought to bear in analyzing the
workings of intertextuality in Web-based public discourse, I now turn to some
examples of texts and their interpretations. The first is an image on the parody site
"False Advertising: A Gallery of Parody" (*False*). The image is of a shipping box with
the familiar "cow" branding logo of the Gateway Computing Company
(Figure 5.1). Above the box are the words "Do it"; on the side of the box,
"GatesWay"; and below the box are the words ". . . or else." The only context
provided for this image in this gallery of images would be the preceding image that
is clearly critical of Microsoft. A user quickly and casually browsing the site might

Figure 5.1. "GatesWay," is used by permission of www.organique.com

conclude that the image is a commentary on Microsoft's proprietary policies regarding use of its products.

A reader who has a broader knowledge of Gateway computing, however, might notice the Gateway logo and wonder what the connection between Gateway and Microsoft CEO Bill Gates might be. But a person with a long-standing knowledge of developments in the tech industry would know that in 1998 Gateway Computing had planned to offer buyers of its computers the choice of using Microsoft's Explorer Web browser or Netscape's Navigator when purchasing its computers (Corrigan). This user would probably also know that Gateway was one of the parties to the antitrust lawsuits that have been filed against Microsoft because of its practice of requiring companies who are less willing to bundle the Explorer browser with their personal computers to pay higher prices for Microsoft software (Kawamoto). The curious user who suspects that there's more behind the image than readily meets the eye might seek out some online news coverage on the topic and discover a detailed background that explains the image.

These reactions to the image illustrate three possible ways of interpreting its context. Readers might simply draw on the preceding image, focus on the juxtaposition of the Gateway logo and the text on the image (parody), or fill in background information from their own knowledge of the larger business and legal contexts (intertextual satire). In addition, an allegory of David and Goliath (the smaller company against Goliath Microsoft) might be at work here. This example illustrates both how different readers can read the same intertextually informed text in different ways depending on their level of knowledge and how different resources can be brought to bear in meaningful interpretations of the text itself. It also illustrates how the reader's past experience and political orientation might affect the intertexts that can be brought to bear to make sense of this parodic image.

The "GatesWay image" worked more or less in isolation; it was one of a set of images posted on a site with no context other than the images that preceded and followed them. It is much more common, however, for verbal, audio, and visual texts to be set in the context of a larger Web site that can itself provide a larger intertextual context for appreciating specific allusions and commentary. For example, in my study of the online Bush versus Gore presidential campaigns in 2000, I discovered that many parody sites worked to set up a larger context within which specific parodic commentaries could be more readily and easily appreciated (*Critical Literacy*). Some of the main parody sites set up facts and content on their home pages (e.g., Bush's malapropisms and Gore's exaggerations) that then functioned as intertexts for later parody and satire. They also generated commonplaces (stock themes, remarks, or sayings) such as "inventor of the internet" or "compassionate conservatism" that could be formulated into texts that subsequently functioned as intertexts. This content was contrived and strategically placed, but it functioned well rhetorically in a communication context designed to appeal to repeat visitors who have been cultivated as in-the-know users and clients of the site.

While most intertextual cross-references are supportive, they can at times be oppositional with regard to other texts in the intertext. By this, I mean that statements and arguments in a given text may be tacitly or explicitly resistive, or refer to critical comments made elsewhere. For this to be effective, readers would need to be familiar with both contexts so as to read the texts in relation to each other. A dramatic instance of this occurred in 2004 when both georgewbush.com and johnkerry.com included regular features on their campaign Web pages that were intended to criticize the opposing candidate or respond to accusations. In their Web site page titled "DBunker" on May 23, 2004, for example, Kerry's site authors quoted a claim on the Bush site that Kerry had voted against tax cuts for the middle class. They then counterposed this claim to a FACT, that "John Kerry has voted repeatedly for marriage penalty relief, including plans specifically designed to ensure that middle class families most penalized by the marriage penalty are the ones who receive the relief" ("DBunker").

In response to such strategies, the Bush site began posting a regular feature, the "DbunkerBuster" link, to respond to Kerry's statements about Bush's policy stances. While some of these charges and countercharges were explicit, others were responses that only visitors to both sites could follow and understand. Evidently, site authors for both sites assumed that readers would be sufficiently interested in the chargers and countercharges to regularly visit both sites (doubtful), or they may have undertaken these Web confrontations so as to influence coverage in the media (more likely).

Jibjab.com: Intertextuality at Work

In July 2004, the animation studio and Web site JibJab posted a short parody, "This Land," intended to lampoon both the Bush and Kerry campaigns. Word of the site spread over the internet through viral dissemination, and it received over 1 million hits in 24 hours (Maney). This parody was the work of two brothers—Gregg and Evan Spiridellis—who wrote, directed, and animated the film. The parody was well received by representatives from both campaigns and by the public because, as one Kerry campaign spokesman said, "it skewers both sides" (Strauss). William Lutz, a commentator on political humor, noted that the animation was "evenhanded and serious about its humor . . . dark and seriously funny" (cited in Strauss). In August 2004, *The Long Island Business News* reported that JibJab had drawn over 10.4 million visitors in July as compared with 3.3 million visitors to the sites of the two presidential candidates (Schachter).

JibJab's success, however, has not been limited to this single parody. It has produced other highly successful works of commentary and parody, including "Big Box Mart," a biting indictment of the marketing and personnel policies of Wal-Mart, and "2-0-5," a commentary on George W. Bush's struggles during his second term in office. The remainder of this section considers a significant factor in the site's rhetorical appeal—its use of intertextuality. By reading the content of two of the site's notable animations in light of their relation to recognizable public events and themes, the verbal and visual texts the parodies draw on, and the message they convey, this section illustrates how JibJab exploits its textual and contextual environments to hold users' attention and influence their thinking.

"This Land Is Your Land"

The animation is divided into six verses. It opens with Bush singing the first verse as he writes place names on a map, spelling Kerry's state "Mass-uh-chew-sits" (Spiridellis and Spiridellis).[6] He's then pictured on a horse in profile with a lasso (emulating the Marlboro Man) and successfully roping Kerry, the "liberal wiener" who is pulled off the screen leaving a Heinz catsup bottle behind. Bush then reminds his listeners that he is a "great crusader" and is pictured in garb similar to Richard the Lion-Hearted with chain mail, crown, white tunic with a cross on it, dancing on a conference table during a meeting of his cabinet.

The second verse is sung by Kerry who notes "your [GWB's] land" is a Texas trailer in the middle of nowhere, and "my [Kerry's] land" is an opulent mansion on well-groomed property. He then notes that he is an intellectual (shown writing complex equations on a blackboard), whereas Bush is a "stupid dumb ass" (shown standing next to the problem "$9 + 3 = 14$" written on the board and a dunce cap on his head). Next, Kerry is shown floating down a river in Vietnam in a swift boat while he displays his three purple hearts pinned inside his camouflage vest and tosses a hand grenade at two Vietnamese villagers on the shore who then are blown to bits.

The third verse positions Bush in front of a House of Pancakes. He reminds listeners that Kerry has "more waffles than a House of Pancakes" and that Kerry offers flip flops whereas Bush offers tax breaks. Bush notes that Kerry is a U. N. "pussy" (and Kerry is positioned in a submissive position with three world leaders (Kofi Annan, Jacques Chirac, and Gerhard Schroeder) standing over him). Bush notes that he can "kick ass" and is pictured riding a bomb down to its target in an image reminiscent of Slim Pickens's exit in *Dr. Strangelove*.

Kerry returns in the fourth verse, noting that Bush cannot say "nuclear" and "that really scares me." While implanting a brain in Bush's head, Kerry notes that "a brain can come in quite handy." But a brain won't help Bush's dropping approval ratings because Kerry has "won three purple hearts." At the end of this verse Kerry is singing along with John Edwards and Howard Dean, who suddenly runs toward the camera screaming.

The two candidates then alternate lines in the next verse, with Bush charging Kerry with being "a liberal sissy," and "a pinko commie" (with hippies, peace signs, and Kerry releasing a dove in the background), while Kerry says that Bush is a "right wing nut job" (photo of Bush sitting in a tank) and "dumb as a doorknob." Although Kerry relies on Botox, he reminds us that he still "won three purple hearts."

The sixth verse opens with a Native American standing in a pristine desert scene saying that "this land was my land" as a nexus of developments and businesses springs up behind him and a chorus sings "but now it's our land." Arnold Schwarzenegger as Rambo and Bill Clinton in his underwear with Monica Lewinsky, being slapped by Hillary then make cameo appearances. The verse concludes with both candidates singing together and saying that "from the liberal wieners" to the "right wing nut jobs" this "land belongs to you and me." The scene then zooms back to show that they are standing together in front of the White House with their backers standing behind them.

Intertextual Commentary on "This Land"

In this animation, the Bush persona characterizes Kerry as a person of privilege, questionable sexual orientation, indecisive, and wimpish, whereas the Kerry persona characterizes Bush as a warmonger, a zealot, and stupid. The parodists also incorporate some commentary on the Republican Party's leanings through their portrayal of corporate greed, and they ridicule Clinton's sexual escapades in this and other animations about the Democrats. A major factor in the success of this specific animation, however, is its effectiveness in using intertextuality as a rich resource for building parody and satire. A rhetorical analysis of the specific cross-references in the animation may enable a better understanding of the reasons why it had the impact that it did.

The animation begins with a play on Bush's many malapropisms and lapses in his knowledge of geography. Examples are legion. Bush once told a group of schoolchildren celebrating "Perseverance Month" that it was important to preserve and that he appreciated their efforts at preservation. Elsewhere, he insisted on the importance of removing the federal "cuff link" that stymies local control of schools (Neal and Romano). He has also maintained that winning the war in Iraq requires an expenditure of money that is "commiserate with keeping a promise to our troops" ("Presidential").

In Jibjab's animated video, Bush begins to spell Kerry's home state and then becomes confused. He is also writing it on the map with a crayon and holding a box of crayons in his hand. To understand the allusion, Jibjab's users must be aware of texts of prior news coverage of George W. Bush's many confusions about geographical locations and the proper use of words. The theme of Bush's stupidity is further spelled out in the later classroom scene when Kerry is writing complex equations on the board whereas Bush, sporting a dunce cap, stands next to his effort: "$9 + 3 = 14$." Among the variations on intertextuality explained earlier in this chapter, these examples would fall under the category of intertextual satire in which satirists try to make a person look ridiculous by drawing on known contexts—in this case, news reports.

In contrast, there are efforts at times to characterize both Bush and Kerry using the second form of cross textual reference described above—references to other specific works and images that are often coordinated to clarify the intertextual references. The Bush persona's appropriation of masculinity and his efforts to emasculate the Kerry persona are interlaced with allusions to past images that are still remembered. These include portrayal of Bush emulating the old advertisements of the Marlboro Man of the Wild West alone on the range. This image of a rugged cowboy in nature with only a cigarette is drawn from a 30-year long advertising campaign for its brand of cigarettes developed by Philip Morris beginning in the 1960s ("Marlboro"). A second image of the Bush persona seated in a tank might remind some users of the photograph of Michael Dukakis seated in an M1 Abrams tank during the 1988 Presidential campaign ("Michael"). He looked so awkward and out of place in the military setting that the photo op became a major public relations disaster for Dukakis's candidacy. For users who recognized this allusion, Bush's role as commander-in-chief at the time of the 2004 election might be contrasted with the earlier Democratic candidate's failure in that role. Finally, the animation's portrayal of Bush riding, cowboy like, on a bomb as he claims to "kick ass," can only bring to mind the aforementioned final scene in the 1964 film *Dr. Strangelove or How I Learned to Stop Worrying and Love the Bomb*. These images convey the idea that the Bush campaign was making a concerted effort to contrast Bush's, the man's man, with Kerry's less aggressive and manly communicative style.

At the same time, Kerry's portrayal as an indecisive wuss and pacificist is supported by references to his marriage to a wealthy woman (aka, the Heinz catsup bottle), which figures prominently in three images—the bondage posture in the scene with U.N. leaders, his former pro-peace stance and fraternization with hippies, and his preference for Botox treatments which, as the sign in the treatment room says, are "not just for old ladies."

Permeating the animation is the song "This Land Is Your Land," which provides the musical narrative to tie the whole together. The song is based on an original version penned by Woody Guthrie who used it to protest class inequality and private ownership. It was rewritten in 1945 with an expanded theme of a "freedom highway" in which the singer roams the land, admiring its vastness and beauty and reminding listeners that the land was made for everyone ("This Land"). The irony of the use of the song is made apparent toward the end of the video, when a Native American man [whose land is it anyway?] stands in an untrammeled desert that is then displaced by corporate franchises such as "Big Buy" [Best Buy], "Mal Mart," "Booger King," and "Nexxo" [Exxon] where a gallon of gas costs $8.95, for regular octane fuel. The animation's use of the Bush persona as a crusader on a quest for a great prize in the land of the infidels along with the juxtaposition with the song "This Land" makes use of allegorical cross-reference to provide acerbic commentary on globalization and the United States' role in it.

The animation as a whole functions as a parody of the respective campaigns' strategies. It also serves as a commentary on the campaigns' inclinations to engage in attack ads and tacitly impugn the character of the opponent. As I have noted, such tendencies themselves were apparent on the Bush and Kerry sites. For example, Souley and Wicks tracked press releases on the two major party presidential campaign sites and found that approximately 80 percent of their online news releases contained at least one reference to the candidates' personality traits. "This Land Is Your Land" thus worked on a meta and a meta-meta level, providing commentary on incidents in the 2004 and prior presidential campaigns (horizontal intertextuality and meta-level commentary) as well as a characterization of the respective candidates' efforts to impugn their opponent's credibility (i.e., Kerry's manliness and Bush's intelligence). Although it is unlikely that all viewers fully appreciated the dozens of intertextual references in the animation, it is very possible that most users appreciated the text in light of the visual and verbal cross-references with which they were most familiar.

"The Drugs I Need"

JibJab's efforts to use its Web site as a venue for public discourse are not limited to political parody. Accompanying "This Land" on its site in 2005 and 2006 was the parody animation "The Drugs I Need," a critique of the pharmaceutical industry and its advertising tactics. To create this film, the country-bluegrass satirists known as the Austin Lounge Lizards collaborated with the Consumers Union by writing the animation's song, and they hired The Animation Farm to create the animation. "The Drugs I Need" was posted on the home page of the Web site "Prescription for Change" (*Prescription*), and subsequently re-posted on JibJab, presumably to increase the animation clip's circulation.

"The Drugs I Need" represents a form of media activism that I will shortly describe when I discuss Adbusters. Media activism often endeavors to focus public attention on media actions that fail to serve the public interest. In an environment in which the American pharmaceutical industry logs over $250 billion annually in sales, where the Food and Drug Administration fails to protect public safety, and where thousands of people have suffered heart attacks when taking the drug Vioxx produced by the Merck company, there is good reason to call attention to drug advertising practices that we see and hear on television and radio (Lieberman).

While many prescribed drugs are undoubtedly beneficial, the marketing and use of Vioxx has led to over 9,000 law suits since it was taken off the market in September 2004 because it was shown to have doubled the risk of heart attacks in a study sponsored by Merck (Henderson). The problem of inadequate review of new drugs is exacerbated when newspaper and television journalists base their reportage about new drugs on drug studies sponsored by the companies themselves, and also when media companies themselves profit greatly from drug product placements and advertising (Lieberman). News reports critical of such practices have noted that drug companies make a regular practice of increasing sales by using advertisements to remind media consumers to take their medicines and also by encouraging physicians to overprescribe medications (Pollack). A recent study of the practices of pharmaceutical firms also concluded that they pursued the practice of "disease mongering," or medicalizing mild conditions such as irritable bowel syndrome and menopause to increase drug sales ("Drug Firms"). Considering the content of "The Drugs I Need" will enable an understanding of the many ways the animation draws on this intertextual context to make its point.

The film opens by reassuring any user who is feeling ill with "some strange disease" not to worry because there's a pill that will set his mind at ease. The pill's name is Progenitorivox, and it's made by "Squabbmerlco." It's "a life-enhancing miracle" but there is a problem; it has some side effects, such as agitation, palpitations, constipation, and the like. "But," the happy pill narrator tells us, "it's worth it for the drugs I need."[7] Even though her disease may not be fatal, a person can ease her fears by "taking two twelve-dollar pills each day for fifty years." Our narrator concludes that he must have Progenitorivox since the pharmaceutical companies have spent billions to convince him that the drug "beats diet and exercise" [here we see a very fat man eating a sandwich while using a fat reduction machine]. The narrator could, of course, buy a generic drug, but he wants Progenitorivox because he "saw it on TV" where "the families look so functional, that paisley pill's for me." Possible outcomes of using Progenitorivox include "deprivation, humiliation, debtors' prison and deportation," not to mention "empty pockets and court dockets." Then the user sees rising in the background a beaver, a moose, and the Canadian flag and is reminded that "IN CANADA THEY GET THIS FOR A SONG" [sung to the tune of the Canadian national anthem]. The animation ends with the customary list of warnings and side effects scrolling across the bottom of the screen, and these include the idea that "unauthorized use of your own judgment in the application of Progenitorivox is strictly prohibited" and a warning that "if death occurs, discontinue use of Progenitorivox immediately" (*JibJab*).

Intertextual Commentary on "The Drugs I Need"

This animation illustrates the extent to which older users and regular consumers of the news are more likely to perceive a larger number of intertextual cross-references than will users who have not been attending to current developments related to the drug industry. The second group is likely to recall frequent television advertisements for drugs that feature an ideal picture of the drug's effects on a person's health and life and simultaneously viewing a routine, rapid-fire list of side effects and warnings shown at the bottom of the screen. The ideality of drug consumption is also connected to other mood-altering drugs and getting "high" as shown in the animation's visual displays of daisies and poppies, content and happy pill-people, and a psychedelic sun shining on the scene. The less-informed user's enjoyment may also be enhanced by visual/verbal, intratextual cross-references generated within the context of the clip, such as flash cards showing people suffering the dysfunctional conditions induced by the drug, the warning messages, and the cartoon photograph of the "functional family" who appear to be on various forms of mood-altering drugs.

A more well-informed viewer of the animation might notice additional connections. The name of the drug "Progenitorivox" has an etymological association with "progeny of Vioxx" and also possibly with Lipitor, the well-known cholesterol-lowering medication. The drug company involved is "Squabbmerlco," a rather transparent allusion to Bristol-Meyers Squibb, makers of Plavix, Pravachol, and other well-known drugs, whose global sales in 2005 were $19.2 billion ("At a Glance"), and to Merck, makers of Vioxx. The phrase about "enhancing life" is taken directly from a motto on Bristol-Myers Squibb's page logo ("At a Glance"). The news-savvy reader will also be aware of the Vioxx lawsuits [court dockets], the significance of the floating dollar bills ["they've spent billions . . ."], and the comparatively lower cost of all drugs in Canada and nearly everywhere else as compared with the United States.

To the extent that the user appreciates a larger number of specific allusions made to other media texts and events, she or he can better understand the message of the animation. This user more or less becomes a coproducer of a coconstructed message. The process is similar to supplying a missing premise when formal reasoning occurs as enthymeme.

It is also interesting to consider the experience of viewing this animation in terms of the semiotic functions described by Lemke. The topic of the film—drug advertising and use—is apparent from its beginning since the theme of pill dependency and the complicity of drug companies are introduced in the first stanza. The user is hailed as someone who may be skeptical about drug dependency, aware of ubiquitous media advertising of drugs, and positioned as exploited by the system. The pill narrator addresses the user as "you" ("you've got a headache . . .") and alludes to his own experience in first person ("I can ease my fears; so now I realize . . ."). The user and the narrator are in the same boat, and the user is invited to sit back, watch, and think about the message of the clip.

The visuals enhance the persuasive effects of the audio effectively, graphically displaying such things as "male lactation," "humiliation," "dire predictions," and "life as seen in Dickens' fictions." Thus, the animation moves from quickly orienting the reader to establishing a connection through shared experience, to organizing the message for the readers' understanding through sequencing events, use of song lyrics, and complementary visuals and audio. Furthermore, by offering both easily understood intratextual references along with more arcane extratextual allusions, the animation can succeed in appealing to a wide range of users with various levels of knowledge of the problem and its background. Since the visitors to the JibJab site are in different age groups and represent a range of political orientations from conservative to liberal, the site takes care to appeal to as many audience orientations as possible.

The differences between its persuasive approach and that of a more activist site might be made clearer if we now turn to Adbusters, an avowedly social activist site, and consider how its appeals to an "in the know" set of users with a specific political orientation might function differently.

Adbusters: Media Activism and Culture Jamming

The mission of the Adbusters Foundation, a loose global network of media activists whose aim is to raise public awareness of the corporatization of media and the negative effects of excessive commercialism, is perhaps best understood through the words of its founder, Kalle Lasn. In a retrospective of his experiences with media practices, Lasn recounted his efforts to place locally produced short films on the Canadian Broadcasting Network. A representative from its sales department told Lasn that they would not air his films despite his willingness to pay for media placement. "I don't know what this is," the man said when viewing his storyboards, "but it's not a commercial" (Lasn 121).

Subsequently, Lasn joined forces with other media activists to try to counter misleading advertisements about the logging industry's forest management practices. Media outlets refused to air the counter ads. The media activists protested, issuing press releases, speaking on talk radio, and staging demonstrations. Following a public outcry, the Canadian Broadcasting Network did not show the counter ads but did discontinue airing the logging industry's ad campaign. Shortly thereafter, Lasn founded The Adbusters Foundation that commenced campaigns such as its annual "Buy Nothing Day," and he began publishing *Adbusters Magazine* and its complementary accompanying Web site. *Adbusters Magazine* began publishing in 1990 and presently has a circulation of about 120,000 ("Adbusters").

A 2002 analysis of the rhetorical appeal and probable impact of Adbusters noted that Lasn's foundation and other media activists such as Subvertise (Subvertise.org) attempt to expose the erosion of public space by commercial interests and to provide a counterpublic where consumerism can be criticized. Rumbo described such organizations as being concerned about "the colonization of public space by marketing and mass media technologies and the degradation of natural environments resulting from global economic growth and concomitant human consumption" (138). Such forms of consumer resistance remind us that discourse about economic conditions can become politicized when activists work to draw public attention to the ubiquitous effects of invasive and exploitive advertising on the public sphere.

Advertising parodies' consumer appeal may be due to the very high level of advertising saturation to which people are continuously exposed. Rumbo noted that consumers in 1996 were being exposed to over twice as many selling messages per day than they were in 1984. Consumers become weary of idealized depictions in ads that leave them feeling that they themselves are "too fat, too ugly, too poor, or the wrong color" (135). Ad parodies reestablish some symmetry between the marketers who control the communication process and consumers; they invite their readers to call up familiar texts in the commercial environment so as to understand and interpret the parody ad. Furthermore, they often require online users to seek out or discover the texts and practices being parodied, thus increasing media literacy (Rumbo). The ways in which this process works can be illustrated by turning to two notable examples of ad parodies on the Adbusters' Web site.

The Benetton Ad

The first of these is a "spoof ad" posted to the Adbusters' Web site ("Spoof"). It pictures a man of about 40 wearing a white shirt and black tie with a wad of cash stuck in his mouth (Figure 5.2). The photo is labeled on the lower right "The True Colors of Benetton." The only colors in the otherwise black and white ad are the man's skin color and the green in the cash and in the label. For the user knowledgeable

Figure 5.2. "The Benetton," are used courtesy of www.adbusters.org

about the advertising industry, this image would probably have immediate meaning. For other users curious about what "Benetton" might be or what it represents, it might have other meanings.

The knowledgeable user would interpret the image as referring to the global, high-end clothing brand, Benetton. This Italian clothing firm is well known for its lines of leisurewear and everyday casual clothing produced in bright colors. The "True Colors of Benetton" in the Adbusters' image is an allusion to the firm's advertising campaigns and their emphasis on "The United Colors of Benetton." Benetton is also well known for its past advertising campaigns that were designed to serve as controversial social statements (Gallagher). These campaigns were originally designed to promote racial harmony and world peace. Early ads featured children and adults of various races interacting with each other. Then they became more controversial and, some would say, sensationalist: photos of a black woman breast-feeding a white baby, a priest kissing a nun on the lips, and a newborn child with the umbilical cord still attached.

A more recent campaign featured death row inmates that caused widespread boycotts of the Benetton project and refusal by Sears and Company to market Benetton products. As one observer noted, "many people were upset not by what they saw, but the context in which it was presented. Critics found it unacceptable that such [ads] were used to drive profits" ("Analysis"). One review of the death row inmates advertisement concluded that "in using actual killers as props for publicizing itself, the clothing company proved sorely lacking in taste and proportion" ("The Lowest").

The *Adbusters* spoof ad, like "This Land," provides a form of meta-level commentary on Benetton's advertising practices. Viewers familiar with Benetton's ads (and many are familiar) will note the deliberate *absence* of color from the spoof ad (in contrast to the pervasive use of bright colors in the Benetton clothesline). The wad of cash in the man's mouth also provides commentary on the very high price of the clothes, and also perhaps on the inconsistency between Benetton's ads' purported focus on progressive issues and its well-recognized profit motive.

In the aftermath of these critical protests and controversy surrounding the sensationalist visuals used in its advertisements, Benetton has developed a different advertising strategy—one more focused on the clothes and intended to emulate the visuals of competitors such as The Gap. Ads currently continue to feature the clothing line's signature bright colors, but they show "exuberant models frolicking in colorful knitwear against a white background" (Gallagher). The appearance of these clothes on slender, fit, teenaged models contrasts sharply with the man in the Adbusters' spoof ad. He appears quite perturbed; they appear serene. He is over 40;

the Benetton models are in their twenties or younger. The only similarity between the Adbusters' image and the images found on the Benetton Web site is the green logo "The True [or] United Colors of Benetton" (*United*).

Thus, the knowledgeable user could draw upon his or her awareness of the development and difficulties of earlier Benetton campaigns to set Adbusters' spoof ad in context. The irony of the company's efforts to strive ostensibly for worldwide brotherhood and world peace while promoting its very expensive line of clothing would not be lost on such a user. Users less knowledgeable about the advertising industry, however, might seek out the Benetton Web site and there might recognize and appreciate the irony of contrast between its sleek ads and the image posted on the Adbusters' site.

The Blackspot Campaign

In Autumn 2004, Adbusters' Kalle Lasn announced a new kind of Web-based campaign. To raise money to support the culture jamming activities of a proposed Blackspot Anti-Corporation, Lasn announced plans to introduce a new product in the athletic shoe market—the Blackspot sneaker (Figure 5.3). Designed to resemble the Converse trainer, acquired by Nike in 2003, the high top version of the Blackspot boasted a sole made of recycled tire material, an organic hemp upper, a hand-drawn antilogo, and a small red dot on the toe intended for "kicking corporate ass." Proceeds from sales of the shoe were to be designated to be spent on "social campaigns and anti-corporate marketing" ("About the Shoes"). Sold primarily online for approximately $45 per pair, about 10,000 pairs of the shoes were ordered within two months after they were introduced ("The Blackspot").

Not only did the Blackspot sneaker produce funds to support the anticorporation, but it also provided a running commentary on the Nike Corporation.

Figure 5.3. "The Blackspot Sneaker," are used courtesy of www.adbusters.org

Nike has historically exploited cheap labor in countries such as Indonesia, China, and Vietnam, where its shoes are made by nonunion workers who are paid 4 dollars per day to make shoes that cost 150 dollars per pair ("Sweatshops"). Although Nike has improved its labor practices and replaced some of the hazardous chemicals used to make its shoes, its workers are still exposed to toxic solvents, adhesives, and other compounds. The Blackspot sneaker, on the other hand, is "made in a safe, comfortable union factory with environmentally sound vegetarian materials" ("About the Shoes"). Furthermore, shoe production occurs at a factory in Portugal with safe working conditions and salaries substantially above minimum wage.

The Blackspot campaign may not have sold a lot of shoes, but it has raised funds for Adbusters' culture jamming activities, increased public awareness of Nike Corporation's exploitive practices, and received a substantial amount of media coverage internationally (e.g., Cunningham; Nolan; Skelton). This campaign illustrates what a social activist media professional can accomplish when he or she combines a Web presence with a media campaign, a message, and an intertext that the campaign's audience understands and appreciates.

Conclusion

Intertextuality's major rhetorical benefit comes from its use of resources in the larger intertext to involve the user in the construction of the text's meaning. In some instances, intertextual references function in the same way as enthymemes did in Aristotle's rhetorical logic. Orators historically have used enthymemes as a form of artistic proof; they have involved their audiences in persuasion by drawing on existing premises and special topics known in the host culture to construct their arguments (Kennedy, *Aristotle*). When Aristotle said that "rhetoric . . . is an ability, in each [particular] case, to see the available means of persuasion," one of the "means" he was referring to was the orator's familiarity with known and accepted probabilities that furnished the materials for proof in support of rhetorical arguments.

In a more or less similar way, authors of online commentary, parody, and satire rely on familiar events, known texts, culturally specific allegories, and other components of the cultural intertext to produce discourses meaningful to various audiences. I say "more or less similar" because, unlike audiences of oratory in Aristotle's time, contemporary users of Web-based discourse have at their fingertips resources that enable them to seek out information in the moment in order to more fully understand and appreciate an intertextual reference. Thus, an observation recently made by David Natharius that "the more we know, the more we see" (241) holds true. That is, the more literate the users are about current events, art forms, and cultural commonplaces, the more they will see and understand. But because intertextuality on the Web may pique user interest and curiosity and because the Web itself offers nearly unlimited opportunities for finding information, it could also be the case that the more the users see, the more they will come to know.

6

Conclusion

The theory and case studies put forward in this book have been intended to explain and illustrate the idea that rhetoric—that is, the persuasive dimension of all forms of discourse addressed to audiences—functions as ubiquitously on the World Wide Web as it does in other communication environments. Since the nature of Web-based texts is in many ways very different from that of print texts and monologic speech, many of the models that have been conventionally used by rhetorical critics and analysts will need to be adjusted for the Web environment. As fully explained in Chapter 2, the emergence of the Web has brought with it a communication context that often is focused as much on the reader as on the message author; addresses dispersed audiences of users consuming modular, disaggregated texts; and conveys messages in a nonlinear mode that is differently consumed by various audiences.

These differences between the Web and prior modes of communication, such as print and speech, apply to the study of public discourse only in part, of course. Our communication environment will continue to include persuasion via books, television, newspapers, and face-to-face communication. Nonetheless, it is advisable for rhetorical critics and analysts to consider systematically the changes in the persuasive process introduced by new media. For example, conventional ways of considering argument structure have drawn upon the whole apparatus of linear logic and sequentially based inference. Due to the hypertext structure and modular workings of Web texts, argument in many new media contexts has assumed a different form. As illustrated in Chapter 5, clever uses of intertextuality, puzzle solving, and suggestive allusion work on the Web to imply positions and points of view that encourage users to supply content and piece together conclusions from the clues supplied in the text. These texts work like enthymemes, since users work through the texts by supplying the missing inferences as they go. Both internal and external intertextuality are phenomena that critics interested in Web-based persuasion should consider.

An additional example of how Web texts differ from many prior media texts can be found in the idea of authorship. Nineteenth- and twentieth-century theories of rhetorical influence emphasized the role of a text's author in crafting messages stylistically suited to the audience and milieu of a speech or printed text. As noted in Chapters 1 and 3, texts on the Web are generally not author centered. Instead, they are made up of the efforts of programmers, designers, writers, and planners. With the exception of blogs and individual sites, many major Web sites with substantial audiences are corporately authored. In studying such sites, critics must move away from the idea of the "work" as designed and authored by a single individual to the idea of the "text" as part of a larger system of hyperlinked and coproduced sites (Barthes). For example, many sites contain content that is an amalgam of affordances posted by site planners and designers and content submitted by site users. The Indymedia site discussed in Chapter 3 and the campaign sites MoveOn and georgewbush discussed in Chapter 4 are examples of sites composed by their users as well as their authors. The implication of this for rhetoric, of course, is the need to move more toward reader-centered criticism of texts. How do readers take up texts in different ways based on the possibilities for consumption offered by the text's authors? What dimensions of the text enable this to happen; how are they structured; and how do they work to encourage users' participation in the texts that they read? Such questions were the focus of Chapters 3 and 4, as I explored possible modes of participation and potential user responses to the texts made possible in the texts themselves.

The focus of this book has been on the use of the Web for persuasive communication in political campaigns, activist resistance, and other efforts to raise public awareness of major social and political issues. The artifacts selected for study in my chapters have included online comment forums, Web site texts addressed to users as a group or groups, Web-based multimedia messages, and other documented and archived forms of public discourse. Although mobile telephony, blogging, and use of the Web interface to enable offline mobilization play important roles in campaigns and resistance activity, the focus of this study has been on the Web site text—its structure, design, content, and rhetorical features.

There are three reasons why it is desirable to proceed in this way. First, considering the nature and rhetorical workings of Web text enables an expanded understanding of its role in the distributed and multimediated Web environment. Not only are users influenced by media elements such as hyperlink patterns, display technologies, and design elements, but they are also influenced by the content of what is said and how it is said in the text as written and communicated. For example, close examination of rhetorical strategies deployed in ostensibly interactive features such as campaign blogs, town halls, and carefully controlled online chat in Chapter 4 revealed ways in which online authors shaped the Web text to *seem* interactive while at the same time insuring that it was not. Considering the rhetorical dimensions of intertextuality in online political parody in Chapter 5 enabled an appreciation of the creativity and art of crafting intertextual allusions that allowed specific texts to appeal to different audiences in different ways over time. As has been shown in a prior study of the user effects of Web-based persuasion (Warnick et al., "Effects"), text-based interactivity played a noticeable role in users' ability to recall and consider what was said on a Web site they had viewed previously. Text-based interactivity consists in part of the use of lively, personalized expression, the content of images, and image placement. On political candidate campaign sites, terms of direct address, personalized photographic portrayals, and allusions to family, constituents, and supporters have an effect. Users remember them after they have left the site (Endres and Warnick). Text-based interactivity as a rhetorical dimension of the Web site text has been understudied, perhaps because it is neither new nor the result of technological development, but it should be included in studies of Web site influence.

My second reason for focusing on the Web site text is that this approach enables development of theories and critical approaches of the sort called for at the beginning of this chapter. Regardless of the extent to which changes in communication technology change the form of the medium, persuasion continues to occur and, when it does, it behooves critics to analyze the means by which people are influenced. One of these means is through the content and nature of the entire site text, including design, hyperlink structure, use of multimedia, forms of interactivity, intertextual allusion, and other textual elements. Thus, the three principal chapters of this book were intended, respectively, to develop a conceptualized theory that accounted for distributed online ethos, to extend conceptual explications of interactivity as a dimension of the site text, and to develop a theoretically informed approach to understanding how intertextuality makes it possible for authors of online Flash, video, and audio content to appeal to multiple audiences.

It is important to keep in mind, however, that Web texts are part of a larger communication environment that includes many other means of communication. One dramatic example of this is the use of a confluence of communication technologies and strategies in political campaigning. As Gronbeck and Wiese have argued, the 2004 political campaigns departed from earlier campaign strategies in major ways. These included use of Web sites and database technologies to parse the electorate into groups that were then targeted for special appeals, the development of nuanced appeals for specific voter groups in key precincts, and the use of other mechanisms that I discussed in Chapter 4 such as meet-ups and special gatherings to bring together and mobilize groups of supporters. Gronbeck and Wiese concluded that in the 2004 election, "we are witnessing a profoundly significant communication event in the annals of Western political process; de-massification and, hence, repersonalization of campaign processes" (529). These trends in the 2004 election were caused partly by campaign finance reform and creative exploitation of the Web to reach individual voters, and the question of whether they will continue and expand in the future is an open one. Nonetheless, these developments work alongside major forms of influence especially enjoyed by young people, such as the use of Flash clips, video animations, and altered images as discussed in Chapter 5. Due to their widespread dissemination and very high hit rate (millions of people), it is very likely that these Web texts played a role in the 2004 election outcome. Rhetorical analysis of Web texts and their contents thus complements analyses of how the Web is used to enable offline involvement and mobilization (Foot and Schneider, *Web Campaigning*).

The third reason for focusing on texts and their content is the need for preservation and a sense of the historical trajectory of the Web's development. For example, my own past and current studies of online parody in the 1996, 2000, and 2004 presidential campaigns show a number of progressive changes in how the forms of parody have changed along with a changing Web environment. During the 1996 campaign between Bob Dole and Bill Clinton, parody sites were amateurish and disorganized, including reciprocal links that confused users, crude image alterations, and veiled slurs on the candidates (Warnick, "Appearance"). Four years later and threatened by legal action during the Bush/Gore campaign, parodists were very careful about what they said. They resorted to using juxtaposed media clips and quotes uttered by the candidates themselves to inform users about candidates' past actions, gaffes, and self-contradictory statements (Warnick, *Critical Literacy*). In 2004, Web parodists resorted to Flash and animations as well as the candidates' own utterances and actions in ridiculing them. They were able to attract sizable audiences and have an impact on the campaigns without the risk of legal action against them. Unlike the situation in 2000, Web-based political parody in 2004 had come into its own as an accepted part of the game. This brief genealogy of online parody illustrates the value of description and analysis of existing Web discourse.

Descriptions and assessments of instances of Web-based public discourse are valuable because the challenges involved in archiving the political Web are nearly insuperable. As I explained in Chapter 2, there is no comprehensive or stable archive of Web documents. The nearest phenomenon we have at present is the Internet Archive. Aside from potential losses due to damage to physical facilities where servers are located, storage medium degradation, loss of funding, and other logistical problems, the IA is an unsatisfactory solution to the problem of Web content preservation. Accessing sites from the past ideally would include access to images, multimedia, second- and third-level links, and other content and features seen when we view texts in real time. This is often not possible when accessing the IA because images have not been stored on a server where they could be accessed and loaded. Since most Web sites now use images for link bars and navigation features, the "look" of the page as originally created is not really visible. Internal links usually do not work, and loading times are extremely slow. Furthermore, site versions are cataloged only minimally according to the capture date.

The degraded condition of political campaign sites in the IA can be compared with a carefully constructed archive of the 2002 mid-term elections, made possible through a collaboration between the Library of Congress and WebArchivist.org and funded by Pew Charitable Trusts. There, each candidate site has a catalog index page and archived contents that can be accessed by office type, party affiliation, geographic location, or name. The image files are linked to the pages so that one can see the page as originally created. Top level and frequently second level links work. This is a good example of an archive that enables scholars and researchers to see the sites in their original form, and there are others.[1] Unfortunately, such archives hold only a miniscule proportion of the historical Web. I have explained this situation at length to illustrate why archiving the Web is unlike archiving photographs, television programs, and other forms of electronic analog media.

Unlike analog media, the Web is largely modular and distributed. Since viewing its historical content requires preservation of prior media forms, use of various versions of prior software and media players, and other, now antiquated technologies, we are at risk of losing our knowledge and awareness of past campaigns and movements. Critical and descriptive analyses of these events by scholars can partially address this problem.

In addition to political campaigns and political parody, another form of public persuasion that has been treated in this project includes political activism, resistance, and oppositional discourse. Chapter 1 discussed online strategies of Greenpeace International, Chapter 3 included a lengthy case study of online credibility at the global Indymedia site, and Chapter 5 examined the use of intertextuality in Adbusters, a culture jamming magazine that critiques corporate advertising. There are numerous other resistance and protest sites, such as the anarchist site Infoshop, Amnesty International, The Nation, and others. In an era when corporate control of media contributes to lack of coverage of issues from protesters' and resisters' points of view, the Web offers a space in which social activists can reach their publics and other potentially supportive groups. The Web functions as an essential platform for resistance and organization of protests worldwide. It also furnishes information on events and actions that go unreported in mainstream corporate media. As Owens and Palmer observed in their study of the 1999 Seattle protests against the World Trade Organization, journalists at that time who wanted to cover the anarchists' activist protest could not just go to a rolodex to find individuals for interviews. Instead, they relied on the Infoshop where they could locate individuals because it offered "access to formerly hard-to-find people based on nominations (that is, links) from the activists themselves" (355).

Due to this access to information and people who would otherwise be difficult to locate, the Web provides a means for underground and resistance groups to affect the slant of news coverage of their actions. It is not the only medium with this capacity, but it may be the most convenient one because it is cataloged, easily searchable, and offers ready information to any user who can locate a site focused on the cause or issue in which the user is interested. Owens and Palmer traced a pattern in which news coverage of anarchist sites in 1999 at first expressed outrage at the property damage in Seattle but then became more moderated as journalists talked with the protesters and developers of anarchist sites, thereby developing a better understanding of their motives and views on the issues. As I noted in Chapter 1, the Greenpeace Web site functions as a valuable arm of that organization when it posts photographs of Greenpeace efforts to interfere with whaling practices or arms tests that are subsequently picked up and published in major papers.

Furthermore, the potential to use open source software, wikis, and other nonproprietary, inexpensive technologies to produce content on the Web is attractive to antiglobalization and resistance groups. Such platforms are attractive not only because of their low cost and ease of use but also because they allow for coproduced content, egalitarianism, widespread participation, and high levels of interactivity when compared with what is possible on proprietary mainstream sites. In the global Indymedia site, however, the result of a low-budget enterprise with a flattened hierarchy of management has produced an unstable system that relies on volunteers, donated equipment, and users' ability to submit high quality, reliable content. For other groups, the development of infrastructure to support a site relies on fund raising and user donations, and work devoted to this may take time away from offline activities and protests. Nonetheless, as Owens and Palmer have argued, the World Wide Web "significantly alters the media landscape of protest, giving activists access to a medium that they themselves control" (336). By providing a venue for activists to reach their publics, the Web plays a role in making counter-hegemonic discourse and actions available to public view. In doing so, it provides a variety of public statements on human rights, openness, and free speech along with the reminder that transnational corporations may be more interested in profit-taking than in the public interest.

It is also important to keep in mind that the Web's malleability and capacity for personalization enable activist and resistance groups to adapt to various audiences. Atkinson and Dougherty used a qualitative interview method to gather viewpoints of antiglobalization proponents. They found that, although some audiences ("the reformists") were interested in learning about the implications and ideologies underpinning transnational globalization, others ("the militants") were more interested in taking action. The latter groups were more likely to be involved in protests, demonstrations, boycotts, and other forms of explicit action. They viewed their use of the Web as a means to organize and mobilize such events.

What one learns when studying uses of persuasion in Web environments is that the rhetoric as persuasive "speech" constitutes a significant domain of online expression. In considering the changes wrought by the shift from print-based text and author-centered speech to the fragmented, multimodal forms of expression we see on the World Wide Web, it might be advisable to take a historical perspective. One model for doing so can be found in Walter J. Ong's *Orality and Literacy* where he charts the shifts in verbal expression from public speaking and oral performance in ancient Greek culture through the development of the written word with its emphasis on linear thought, formal logic, and an objective point of view. Ong welcomed the advent of electronic communication and, with it, the emergence of secondary orality with its emphasis on communal experience, participation, and immediacy of expression. He also believed, however, that conventional literacy, the written word, and analysis were essential components of reflective thinking.

The aim of this book has been to revisit forms of Web-based public expression that have this quality of secondary orality described by Ong. "Talk" on the World Wide Web has the quality of unending conversation described by Kenneth Burke in *The Philosophy of Literary Form*, where various constituencies speak their minds, reach out to potential converts and supporters, and seek to mobilize others. These advocates do this through use of persuasive forms of online speech: for example, by presenting themselves in the most favorable light, lending presence to their cause, crafting messages for heterogeneous audiences, and expressing their points of view through parody and satire. Although many of their communications may be discontinuous rather than linear, coproduced rather than author centered, and ephemeral rather than inscribed, rhetoric continues to play a role in forming ideas and shaping attitudes, as it always has. Hopefully, this book will play a role in improving understanding of how this process happens in the ubiquitous environment of the World Wide Web.

Examples of Texts Cited

Chapter 2   Online Rhetoric:
A Medium Theory Approach

Slashdot: News for Nerds. Stuff that Matters. 2006. Open Source Technology Group. Retrieved April 22, 2006 <*http://slashdot.org/*>.

Wikipedia. 2006. Wikipedia Foundation, Inc. Retrieved April 22, 2006 <*http://en. wikipedia.org/wiki/Main_Page*>.

Wilkins, Edo. "State of the Union. . . . Not Good." *Fuckitall.com.* 2000. fuckitall.com. Retrieved April 18, 2006 <*http://fuckitall.com/bsh/*>.

Chapter 3   The Field Dependency of
Online Credibility

IMDb: Earth's Biggest Movie Database. 2006. Internet Movie Database Inc. Retrieved April 24, 2006 <*http://www.imdb.com/*>.

MedlinePlus: Trusted Health Information for You. April 24, 2006. U.S. National Library of Medicine and the National Institutes of Health. Retrieved April 24, 2006 <*http:// medlineplus.gov/*>.

"United States Government Assassinates Leader of Puerto Rican Independence." Retrieved November 14, 2005 <*http://www.indymedia.org/en/2005/09/824994.shtml*>.

Chapter 4   Interactivity: The Golden Fleece of
the Internet

"About Buddy." *Darden for Congress.* 2002. Retrieved August 10, 2002 <*http://www. dardenforcongress.net/muse?Element.1.2.5=page*>.

"Action Center." *GeorgeWBush.com*. n.d. Retrieved December 25, 2005 <*http://web.archive.org/web/20040610111929/www.georgewbush.com/GetActive/*>.

"Attend a Vote for Change Watch Party." *Moveon.Org; Political Action*. n.d. Retrieved December 27, 2005 <*http://web.archive.org/web/20041012022020/ action.moveonpac. org/vfc/*>.

"Bush/Cheney'04." *GeorgeWBush.com*. n.d. Retrieved December 25, 2005 <*http://web. archive.org/web/20041015075713/http://www.georgewbush.com/*>.

"I Support President Bush Because . . ." *GeorgeWBush.com*. n.d. Retrieved December 25, 2005 <*http://web.archive.org/web/20040610175041/www.georgewbush.com/News/ support.aspx*>.

"Kerry Media Center." *GeorgeWBush.com*. October 14, 2004. Retrieved December 25, 2005 <*http://web.archive.org/web/20041015080558/www.georgewbush.com/Kerry MediaCenter/*>.

"Official Campaign Blog." *GeorgeWBush.com*. October 14, 2004. Retrieved December 25, 2005 <*http://web.archive.org/web/20041015052317/http://georgewbush.com/blog/*>.

"Other Question or Problem." *MoveOn.org: Political Action*. n.d. Retrieved December 25, 2005 from link from <*http://political.moveon.org/feedback/fb/index. html?entity=pac*>.

"Stan the Man." *Stan Matsunaka for Congress*. Posted July 21, 2002. Retrieved December 24, 2005 from the IA at <*http://web.archive.org/web/20020721075309/ http://www.stan 2002.com/bio.html*>.

"Recent Success Stories." *MoveOn.org: Democracy in Action*. n.d. Retrieved December 24, 2005 <*http://www.moveon.org/success_stories.html*>.

"Vigils to Honor 2,000 Killed in Iraq." *MoveOn.org: Political Action*. n.d. Retrieved December 24, 2005 <*http://political.moveon.org/iraqvigils/*>.

"Why We Did This." *Bush in 30 Seconds: A Political Advertising Contest Sponsored by Moveon.org Voter Fund*. n.d. Retrieved December 24, 2005 <*http://www.bushin30 seconds.org/aboutus.html*>.

Chapter 5   Intertextuality and Web-Based Public Discourse

Apple Macintosh Computer. Advertisement. n.d. http://www.uriahcarpenter.info/. Retrieved March 20, 2006 <*http://www.uriahcarpenter.info/1984.html*>.

"At a Glance." *Bristol-Myers Squibb Company*. 2005. Bristol-Myers Squibb Company. Retrieved April 8, 2006 <*http://www.bms.com/aboutbms/data/*>.

"Truthful Translations of Political Speech or, 'What They Really Meant When They Said That.' " *DiYMedia.net*. 2006. DiYMedia. Retrieved July 1, 2006 <*http://www.diymedia. net/collage/truth.htm*>.

False Advertising: A Gallery of Parody. n.d. Retrieved March 27, 2006 <*http://parody. organique.com/024_1.html*>.

JibJab. 2005. Retrieved April 8, 2006 <*http://www.jibjab.com/Home.aspx*>.

"More Google: Holiday Logos." *Google.com*. 2006. Google Inc. Retrieved April 14, 2006 <*http://www.google.com.offcampus.lib.washington.edu/holidaylogos.html*>.

Prescription for Change. 2003. Consumers Union. Retrieved April 8, 2006 <*https://secure.npsite.org/cu/site/Advocacy?page=UserAction&cmd=display&id=691*>.

Spiridellis, Gregg and Evan Spiridellis. "This Land." *JibJab*. 2004. JibJab Media Inc. Retrieved April 10, 2006 <*http://www.jibjab.com/Home.aspx*>.

"Spoof Ads: Fashion." *Adbusters*. n.d. Adbusters Media Foundation. Retrieved April 12, 2006 <*http://adbusters.org/spoofads/fashion/benetton/*>.

Subvertise.org. 2001. www.subvertise.org. Retrieved April 12, 2006 <*http://subvertise.net/contact.php*>.

"The Drugs I Need." *JibJab*. JibJab Media Inc. Retrieved April 10, 2006 <*http://www.jibjab.com/Home.aspx*>.

Wilkins, Edo. "State of the Union . . . Not Good." *Fuckitall.com*. 2000. fuckitall.com. Retrieved April 18, 2006 <*http://fuckitall.com/bsh/*>.

Conclusion

Amnesty International. 2006. Retrieved April 29, 2006 <*http://www.amnesty.org./*>.

Infoshop.org. Ed. anonymous. 2006. Retrieved April 29, 2006 <*http://www.infoshop.org/*>.

The Nation. Ed. Katrina V. Heuvel. 2006. The Nation. Retrieved April 29, 2006 <*http://www.thenation.com/*>.

Notes

Chapter 1   The Internet and the Public Sphere

1. See Habermas, Calhoun, and Castells for further discussion of these characteristics.
2. Bruner described the intended purposes of these organizations:

   The goal of the World Bank was to facilitate the reconstruction of Europe and then the "developing" countries through grants and loans. The goal of GATT (General Agreement on Tariffs and Trade, which later became the World Trade Organization) was to multilaterally lower trade barriers, manage global trade policy, and prevent dependent member countries from raising tariffs, providing domestic subsidies, or otherwise tampering with free and open trade. The IMF's goal was to provide stability in international currencies and international exchange. (693)

3. These include news talk shows such as *Larry King Live* and Sunday interview programs, televised candidate appearances, and televised debates.
4. The most recent report by the Pew Internet and American Life Project (Fox) as of this writing indicated that both African Americans and English-speaking Latinos had made substantial gains on whites in Internet access.
5. Sources that provide accounts of such actions include Castells, Kellner, and collected essays in McCaughey and Ayers and in Schuler and Day.
6. As I explain later in this chapter, blogs (or Web logs) are online journals where individuals record their personal experiences, thoughts, and opinions.
7. Television and newspaper news coverage also have altered their visual and presentation formats to emulate the Web, perhaps because of the need to appeal to users accustomed to quickly consumed, segmented, multimodal displays. For example, it is not uncommon to view a televised news anchor in front of projected multiple images of an ongoing crisis. My local newspaper (*The Seattle Post-Intelligencer*) recently reformatted its international section to include "The World in Five Minutes," a horizontal display across

two pages of syndicated photos plus captions (but no story) of events around the world. (These items appear to have been selected for their visual appeal, not their news value.)

8. In the 2000 election, Zack Exley, a 29-year-old computer programmer from New York, posted a Web site lampooning George W. Bush. When the Bush campaign complained to the Federal Election Commission and tried to block the site, the story was covered by *The Washington Post*, *The New York Times*, and the *Los Angeles Times*. Internet traffic on Exley's site then increased from a trickle to over a million visitors in 1999. I have described these developments in my book *Critical Literacy*.

9. This work has appeared in the *Journal of Computer-Mediated Communication*, the *Electronic Journal of Communication*, *Critical Studies in Media Communication*, and in a number of anthologies.

10. A review of work in this area prior to 2001 appeared in my book, *Critical Literacy*. To identify noncritical rhetorical analyses of the uses of the Internet for social activism, I searched the 2001–2005 issues of the *Quarterly Journal of Speech, Rhetoric and Public Affairs, Critical Studies in Media Communication*, and three regional communication journals but did not locate any examples.

Chapter 2 Online Rhetoric: A Medium Theory Approach

1. I say "usually" because authorship of spoken and print texts is often a corporate enterprise involving ghost writers, speechwriters, collaborators, translators, and other contributors to the final product. Some portion of what we read has no named author. Due to the technological and design requirements for Web site development, however, corporate authorship in many online genres has become the rule rather than the exception.

2. It may be premature and speculative to chart trends in literacy levels and attribute them to the rising use of the internet as a primary cause. There is some indication, however, that there have been major changes in reading habits and patterns of media consumption. For example, a 2002 large-scale survey of 17,000 individuals conducted by the Census Bureau concerning public participation in the arts indicated that the percentage of Americans reading literature has declined from 56.9 in 1982 to 46.7 in 2002, and that there has been a decrease in the percentage of respondents saying that they had "read any book" of over 7 percent between 1992 and 2002. The study noted that "during the time period when literature participation rates declined, home Internet use soared" (from 28 percent in 1998 to 42 percent of households in 2002) (National Endowment 30). The study was cautious in interpreting these findings, since other causes may account for changes in reading patterns, but it calls for more research on the public's reading habits and their relationships to online activity. With regard to newspaper reading specifically, a *New York Times* article in October 2005 noted that two-thirds of U.S. households are expected to have high-speed internet connections by 2010 and that newspapers are losing advertising revenue as

shoppers move online to buy goods. As with books, print newspaper readership is declining. While 52 percent of adults read a newspaper every day, these readers are generally in the older demographic groups. In the last 25 years, the percentage of people aged 30–39 who read a newspaper every day had dropped from 73 percent to 30 percent (Seelye). The problem of unequal access to technology has been placed under the rubric "the digital divide" and, during the 1990s, focused on differential access as related to gender and race. More recently, since nearly everyone interested in being online has achieved access, concern about the digital divide has shifted to other factors. These include differences in quality of hardware and connection speed due to income and affordability as well as factors such as literacy levels, education, technological proficiency, and geographic location. While the advent of wireless access technologies and mobile telephony has made access to the World Wide Web more widespread, these technologies are still under development (Souheil and Blanchard).

3. Ott and Walter noted that feminist critics such as Julia Kristeva and media scholars each use the term "intertextuality" to refer to two distinct phenomena. The first form of intertextuality refers to an interpretive practice unconsciously used by audiences—a sensibility that conceives of texts as fragments in a larger web of textuality. The second refers to stylistic devices consciously used to make specific lateral associations between texts. It is this second form of lateral reference that is discussed in Chapter 5 of this book.

Chapter 3   The Field Dependency of Online Credibility

1. Explanation of users' evaluations of Web site credibility as identified in empirical studies is discussed in the next section on "Research Findings."
2. Wathen's and Burkell's account of how users respond to Web sites was presented by them as an inclusive account, but it was in fact based on studies of people using sites that were "institutionalized" in some sense, such as commercial, educational, nonprofit, or government sites. Users' reactions to noninstitutionalized sites such as blogs, activist sites, and coproduced sites might be quite different.
3. There are two remaining components in the Toulmin model that I did not discuss here—reservation and qualifier. The reservation impacts the evidence/claim link by stating instances in which that link would not hold up. In the example below about the Internet Movie Database Web site, a reservation could be stated that "the IMDB seems generally reliable [claim] so long as its movie reviewers are as well informed as the site implies that they are." The reservation, then, can be used to qualify the cogency of the claim. The qualifier in Toulmin's model indicates the degree of likelihood or cogency of the claim and is usually expressed as an adverb modifying the claim statement, such as "probably," "generally," or "in most cases."
4. Research on the quality of information on medical health sites has clearly indicated the importance of these criteria. See Kunst et al.; Waljii et al.; and Zun et al. for examples.

Chapter 4   Interactivity: The Golden Fleece of the Internet

1. "Interactivity" is labeled as a "golden fleece" in Burnett and Marshall (76), and elsewhere in the literature.

2. A notable exception to the trend in low interactivity between users on political campaign sites was the Dean for America campaign in 2003 and early 2004. This Web-based campaign, managed by Joe Trippi, was designed to brand Dean as the candidate who would use the internet to revive democracy. It included the campaign's alignment with MeetUp.com to bring together Dean supporters in face-to-face meetings throughout the United States, an open blog in which over 300,000 comments from users were posted, and efforts by online supporters to suggest innovative person-to-person campaign strategies (Cornfield; Haas).

3. In a 2005 study of user responses to specific dimensions of interactivity in political candidate sites (Warnick et al.), users were asked to provide open-ended explanations for their responses to features on the sites. They specifically remembered examples of loaded language and negative campaigning in the site text. Furthermore, they also commented favorably on the candidate's family orientation as shown in photographs displayed on the site.

4. See an October 2004 version of the site at <http://web.archive.org/web/20041009122012/www.moveon.org/front/> and a June 2004 version at <http://web.archive.org/web/20040610002454/www.moveon.org/front/>.

5. A view of MoveOn's home page in December 2005 was retrieved on December 27, 2005 from <http://www.moveon.org/> (It is not possible to provide durable links for site versions that have not yet been archived; ultimately these may be found by entering the site's URL at the time of original access into the search box at the IA. Capture dates will be displayed, and then the user should select the access date closest to the retrieval date.)

6. Georgewbush.com was selected for this study because it offered an example of a major presidential candidate campaign site in 2004 that contrasted in strategy with the approaches taken by MoveOn Civic Action (a nonprofit organization) and MoveOn Political Action (a federal PAC). I had originally planned to compare two sites in the same category—Bush's and Kerry's sites—but I decided that more was to be gained by comparing two political sites in different categories. In addition, archived versions of Kerry's site during 2004 became suddenly inaccessible in late 2005 when I needed to access them because of technical problems with the IA. Mr. Paul Hickman, an office manager at the archive, emailed me the following message on December 6, 2005: "Thank you for contacting the Internet Archive. I apologize for the delay in responding to your request. We are having some server issues, and some sites are having some issues or not appearing at all. Please be patient, and we are working on it as quickly as possible" (Hickman). It is

important to reemphasize the fragility of preservation of Internet content that I discussed in Chapter 2. The archiving of a presidential campaign is a good test case of how well we can preserve our online political history. Whereas the 2004 versions of John Kerry reappeared in late December and can be located at <http://web.archive.org/web/*/http://www.johnkerry. com>, this experience is an indication of how unstable the preservation situation currently is.

7. The georgewbush.com site design did not change appreciably during 2004, so the version evaluated by this group of designers strongly resembled the design as it appeared throughout the campaign season. For my analysis of interactivity on the georgewbush site in 2004, I compared site captures from October 15, 2003; January 26, 2004; June 4, 2004; and October 15, 2004. For the January version at <http://web.archive.org/web/ 20040126055651/http://georgewbush.com/> and the October version at <http://web.archive.org/web/20041015075713/ http://www. georgewbush.com/>, 23 primary links on the left tool bar and across the top of the site were the same, and 12 were different. The design was also essentially the same, with multiple navigation bars.

8. In an article about this particular online chat with the Bush daughters, Kever noted that about 2,600 questions were submitted by users, and the twins answered 14 selected questions.

9. For the archived June version of the georgewbush site, see <http://web.archive.org/web/20040604163152/http://www.georgewbush.com/>; for the October version, see <http://web.archive.org/web/20041015075713/ http://www.georgewbush.com/>.

Chapter 5   Intertextuality and Web-Based Public Discourse

1. Material for the preceding three paragraphs was contributed by Timothy Pasch.

2. For example, in the 2000 election, the Web site Algore2000.org posted two clips from Gore's speeches. One, to the 1996 Democratic convention, was about his sister's illness and death from cancer caused by smoking, and the other, to tobacco farmers in Tennessee after his sister's death, described Gore's pride in helping out on his family's tobacco farm. See Warnick, *Critical Literacy*, 102.

3. Examples include the widely circulated image of John Kerry as Herman Munster that appeared in the JibJab animation, "This Land" (Spiridellis and Spiridellis) and a patched-in version of George W. Bush's State of the Union message in 2000 on the site fuckitall.com (Wilkins).

4. This and all subsequent quotations from portions of this Flash parody were transcribed from what was said on the site.

5. The way in which the user is addressed (e.g., use of pronouns) is also an important factor in text-based interactivity as discussed in Chapter 4.
6. All citations to "This Land" in this section are directly from the video posted to JibJab and produced by Spiridellis and Spiridellis.
7. All citations to "The Drugs I Need" in this section are to the lyrics and displayed text in the animation by this title posted on the JibJab site.

Chapter 6   Conclusion

1. See, for example, the September 11 Digital Archive at <http://www.911digitalarchive.org/>, the hurricane digital memory bank at <http://www.hurricanearchive.org/index.php>, and the Indymedia archive at <http://www.indymedia.org/en/feature/archive.shtml>.

Works Cited

"2004 Presidential Election." *Opensecrets.org*. n.d. Retrieved December 25, 2005 <*http://www.opensecrets.org/presidential/index.asp?sort=R*>.

Aarseth, Espen J. *Cybertext: Perspectives on Ergodic Literature*. Baltimore: Johns Hopkins University Press, 1997.

"About Copyscape." *Copyscape*. 2006. Indigo Stream Technologies. Retrieved April 7, 2006 <*http://copyscape.com/about.php*>.

"About Indymedia." Retrieved November 23, 2005 <*http://www.indymedia.org/en/static/about.shtml*>.

"About the Internet Archive." *Internet Archive*. Retrieved October 27, 2005 <*http://www.archive.org/about/about.php*>.

"About the MoveOn." *MoveOn.org: Democracy in Action*. n.d. Retrieved December 24, 2005 <*http://www.moveon.org/about.html*>.

"About the Shoes." *Blackspot Shoes*. n.d. Adbusters.org. Retrieved April 14, 2006 <*http://adbusters.org/metas/corpo/blackspotshoes/info.php*>.

"Acid Software Overview." *Acid Planet*. 2006. Sony Corporation. Retrieved April 1, 2006 <*http://www.acidplanet.com/tools/?p=acid&T=4692*>.

"Adbusters." *Wikipedia*. 2006. Retrieved April 10, 2006 <*http://en.wikipedia.org/wiki/AdBusters*>.

"Allegory." *Cambridge Dictionaries Online*. 2006. University of Washington. Retrieved March 21, 2006 <*http://dictionary.cambridge.org.offcampus.lib.washington.edu/define.asp?key=2168&dict=CALD*>.

"Allegory." *Encyclopedia Britannica Online*. 2006. University of Washington. Retrieved March 22, 2006 <*http://www.search.eb.com.offcampus.lib.washington.edu/eb/article-9005781?query=allegory&ct=eb*>.

Allen, Graham. *Intertextuality*. London: Routledge, 2000.

Althusser, Louis. "Ideology and Ideological State Apparatuses (Notes toward an Investigation." *Lenin and Philosophy, and Other Essays*. Trans. B. Brewster. New York: Monthly Review Press, 1972. 127–186.

American Psychological Association. *APA Style.org: Electronic References*. Retrieved September 1, 2005. <*http://www.apastyle.org/electext.html*>.

"Analysis: Should Benetton Water Down Its Advertising Impact?" *Marketing Week* (September 22, 2005): 15. *ProQuest*. University of Washington. Retrieved April 12, 2006 <*http://www.proquest.com*>.

Aristotle. *On Rhetoric: A Theory of Civic Discourse*. Trans. George A. Kennedy. Oxford: Oxford University Press, 1991.

Atkinson, Joshua and Debbie S. Dougherty. "Alternative Media and Social Justice Movements: The Development of a Resistance Performance Paradigm of Audience Analysis." *Western Journal of Communication* 70.1 (2006): 64–88.

Atton, Chris. *Alternative Media*. London: Sage, 2002.

Atton, Chris. "Indymedia and 'Enduring Freedom': An Exploration of Sources, Perspectives and 'News' in an Alternative Internet Project." In *Studies in Terrorism: Media Scholarship and the Enigma of Terror*. Ed. N. Chitty, R. R. Rush, and M. Semati. Penang, Malaysia: Southbound, 2003. 147–164.

Atton, Chris and Nick Couldry. "Introduction." *Media, Culture & Society* 25 (2003): 579–586. Sage. Retrieved November 23, 2005 <*http://www.sagepub.co.uk/Journalhome.aspx*>.

Bakhtin, Mikhail M. *The Dialogic Imagination: Four Essays*. Trans. Caryl Emerson and Michael Holquist. Austin: University of Texas Press, 1981.

Ballve, Teo. "Another Media Is Possible." *NCLA Report on the Americas* 37.4 (2004): 29: 29. Retrieved July 29, 2005 <*http://proquest.umi.com*>.

Bancroft, Collette. "Political Ads Go Pop." *St. Petersburg Times* (December 26, 2003): 1E. Retrieved December 13, 2005 from *Lexis Nexis* <*http://www.lexisnexis.com/*>.

Barthes, Roland. *Image Music Text*. Trans. Stephen Heath. New York: The Noonday Press, 1977.

Barton, Matthew D. "The Future of Rational-Critical Debate in Online Public Spheres." *Computers and Communication* 22 (2005): 177–190. Elsevier. Retrieved June 22, 2005 <*www.sciencedirect.com*>.

Baumlin, James. S. "Ethos." In *Encyclopedia of Rhetoric*. Ed. Thomas O. Sloane. Oxford: Oxford University Press, 2001.

Beckerman, Gal. "Edging Away from Anarchy: Inside the Indymedia Collective, Passion vs. Pragmatism." *Columbia Journalism Review* 42.3 (2003): 27–30.

Bennett, W. Lance. "Toward a Theory of Press-State Relations in the United States." *Journal of Communication* 40.2 (1990): 103–125.

Bimber, Bruce and Richard Davis. *Campaigning Online: The Internet in U. S. Elections*. Oxford: Oxford University Press, 2003.

Bizzell, Patricia and Bruce Herzberg, eds. *The Rhetorical Tradition: Readings from Classical Times to the Present*. Boston: Bedford St. Martin's, 1990.

"The Blackspot Sneaker." *Intheblack* 74.11 (2004): 14+. *ProQuest*. University of Washington. Retrieved April 14, 2006 <*http://www.proquest.com*>.

Bolter, Jay David. "Theory and Practice in New Media Studies." In *Digital Media Revisited*. Ed. Gunnar Liestøl, Andrew Morrison, and Terje Rasmussen. Cambridge: The MIT Press, 2003. 15–33.

Bolter, Jay David and Diane Gromala. *Windows and Mirrors: Interaction Design, Digital Art, and the Myth of Transparency*. Cambridge: The MIT Press, 2003.

Bostdorff, Denise M. "The Internet Rhetoric of the Ku Klux Klan: A Case Study in Web Site Community Building Run Amok." *Communication Studies* 55 (2004): 340–361.

Boyd, Josh. "The Rhetorical Construction of Trust Online." *Communication Theory* 13.4 (2003): 392–410.

Brown, Richard Harvey. "Global Capitalism, National Sovereignty, and the Decline of Democratic Space." *Rhetoric and Public Affairs* 5 (2002): 347–357.

Bruner, M. Lane. "Global Governance and the Critical Public." *Rhetoric and Public Affairs* 6 (2003): 687–708.

Bucy, Erik P. "Interactivity in Society: Locating an Elusive Concept." *The Information Society* 20 (2004): 373–383.

Burbules, Nicholas C. "Paradoxes of the Web: The Ethical Dimensions of Credibility." *Library Trends* 49.3 (2001): 441–455.

Burke, Kenneth. *A Rhetoric of Motives*. Berkeley: University of California Press, 1969 [1950].

Burke, Kenneth. *The Philosophy of Literary Form*. 3rd edn. Berkeley: University of California Press, 1973.

Burnett, Robert and P. David Marshall. *Web Theory: An Introduction*. London: Routledge, 2003.

Bury, Rhiannon. "Language on the Line: Class, Community and the David Duchovny Estrogen Brigades." *Electronic Journal of Communication* 14 (2004). Communication Institute for Online Scholarship. Retrieved October 27, 2005 <*http://www.cios.org/www/ejcmain.htm*>.

Calhoun, Craig. "Information Technology and the International Public Sphere." In *Shaping the Network Society: The New Role of Civil Society in Cyberspace*. Ed. Douglas Schuler and Peter Day. Cambridge: The MIT Press, 2004. 229–251.

Carey, James W. "Historical Pragmatism and the Internet." *New Media & Society* 7.4 (2005): 443–455.

Castells, Manuel. *The Power of Identity*. 2nd edn. Malden: Blackwell, 2004. Vol. 2 of *The Information Age: Economy, Society, and Culture*. 3 vols. to date 1996– .

Castells, Manuel. *The Rise of the Network Society*. Malden: Blackwell, 1996. Vol. 1 of *The Information Age: Economy, Society, and Culture*. 3 vols. to date 1996– .

Ceccarelli, Leah. "Polysemy: Multiple Meanings in Rhetorical Criticism." *Quarterly Journal of Speech* 84 (1998): 395–415.

Cha, Ariana Eunjung. "Grass-Roots Politics with Click of a Mouse; In Silicon Valley, Tech-Driven Support Groups." *The Washington Post* (October 25, 2004), final edn.: A03. Retrieved December 31, 2005 from *Lexis Nexis* <*http://www.lexisnexis.com/*>.

Chen, Sherry Y. and Robert D. Macredie. "Cognitive Styles and Hypermedia Navigation: Development of a Learning Model." *Journal of the American Society for Information Science and Technology* 53 (2002): 3–15.

Consumers International. "Credibility on the Web: An International Study of the Credibility of Consumer Information on the Internet." 2002. Retrieved May 9, 2004 <*http://www.consumersinternational.org/Publications/search.asp?langid=1&regid=135*>.

Consumers International. "Credibility on the Web: An International Study of the Credibility of Consumer Information on the Internet." 2002. Retrieved October 27, 2005 <*http://www.consumersinternational.org*>.

Cooke, Lynne. "A Visual Convergence of Print, Television, and the Internet: Charting 40 Years of Design Change in News Presentation." *New Media & Society* 7.1 (2005): 22–46.

Cornfield, Michael. "The Internet and Campaign 2004: A Look Back at the Campaigners." Retrieved December 24, 2005 <*http://www.pewinternet.org/*>.

Cornfield, Michael, Shabbir Safdar, and Jonah Seiger. "The Top 10 Things We're Tired of Seeing in Candidate Web Sites." *Campaigns & Elections* 19.9 (1998): 26.

Corrigan, Louis. "Gateway Wins One against Gates." *The Motley Fool.* May 28, 1998. Retrieved March 27, 2006 <*http://www.fool.com/LunchNews/1998/LunchNews980528.htm*>.

Coyle, James R. and Esther Thorson. "The Effects of Progressive Levels of Interactivity and Vividness in Web Marketing Sites." *Journal of Advertising* 30.3 (2001): 65–77. Retrieved December 24, 2005 from Infotrac at <*http://infotrac.galegroup.com*>.

Cunningham, Jennifer. "A Curse on All Brands: The Blackspot Sneaker Is Determined to Give Global Giant Nike a Kicking." *The Herald (Glasgow (UK)* (September 29, 2004): 13. *ProQuest.* University of Washington. Retrieved April 14, 2006 <*http://www.proquest.com*>.

"Dbunker: Setting the Record Straight." *John Kerry for President.* June 17, 2004. johnkerry.com. Retrieved March 27, 2006 <*http://web.archive.org/web/20040619055559/blog.johnkerry.com/dbunker/*>.

Downey, John and Natalie Fenton. "New Media, Counter Publicity and the Public Sphere. *New Media & Society* 5 (2003): 185–202. Sage. Retrieved June 13, 2005 <*http://nms.sagepub.com*>.

Downing, John D. H. "Audiences and Readers of Alternative Media: The Absent Lure of the Virtually Unknown." *Media, Culture & Society* 25 (2003): 625–645. Sage. Retrieved November 22, 2005 <*http://www.sagepub.com/*>.

"Drug Firms 'Inventing Diseases.'" *BBC News.* April 11, 2006; April 17, 2006 <*http://news.bbc.co.uk/2/hi/health/4898488.stm*>.

Eaton, Phoebe. "On the Web, an Amateur Audience Creates Anti-Bush Ads." *The New York Times* (December 21, 2003): 6. Retrieved December 13, 2005 from *Lexis Nexis* <*http://www.lexisnexis.com/*>.

Edwards, Eli. "Ephemeral to Enduring: The Internet Archive and Its Role in Preserving Digital Media." *Information Technology and Libraries* 23.1 (2004). *Proquest.* Retrieved October 11, 2005 http://www.proquest.com.

Emigh, Emil and Susan C. Herring. "Collaborative Authoring on the Web: A Genre Analysis of Online Encyclopedias." *Proceedings of the Thirty-Eighth Hawai'i International Conference on System Sciences (HICSS-38).* Los Alamitos: IEEE Press, 2005. Retrieved August 29, 2005 <*http://ella.slis.indiana.edu/~herring/wiki.pdf*>.

Endres, Danielle and Barbara Warnick. "Text-Based Interactivity in Candidate Campaign Web Sites: A Case Study from the 2002 Elections." *Western Journal of Communication* 68 (2004): 322–342.

Enos, Theresa and Shane Borrowman. "Authority and Credibility: Classical Rhetoric, the Internet, and the Teaching of Techno-Ethos." In *Alternative Rhetorics: Challenges to the Rhetorical Tradition.* Ed. Laura Gray Rosendale and Sibyle Gruber. Albany: State University of New York Press, 2001. 93–109.

Entman, Robert E. "Framing U.S. Coverage of International News: Contrasts in Narratives of the KAL and Iran Air Incidents." *Journal of Communication* 41.4 (1991): 6–27.

Fagerjord, Anders. "Rhetorical Convergence: Studying Web Media." In *Digital Media Revisited.* Ed. Gunnar Liestøl, Andrew Morrison, and Terje Rasmussen. Cambridge: The MIT Press, 2003. 293–325.

Farkas, David K. and Jean B. Farkas. *Principles of Web Design.* New York: Longman, 2002.

Fattah, Hassan M. "Man in Infamous Photo Tells of Abu Ghraib Ordeal." *Seattle Post-Intelligencer* (March 11, 2006): A4.

Fiske, John. *Television Culture.* London: Methuen, 1987.

Fogg, B. J., Cathy Soohoo, and David Danielson. *How Do People Evaluate a Web Site's Credibility? Results from a Large Study.* 2002. Retrieved October 27, 2005 from Stanford Persuasive Technology Lab on the Consumer WebWatch site: <*http://www.consumerwebwatch.org/view-article.cfm?id=10048&at=510*>.

Foot, Kirsten A. and Steven M. Schneider. "Online Structure for Civic Engagement in the September 11 Web Sphere." *Electronic Journal of Communication* 14 (2004). August 25, 2005 <*http://www.cios.org/www/ejcmain.htm*>.

Foot, Kirsten A. and Steven M. Schneider. *Web Campaigning.* Cambridge: The MIT Press, 2006.

Foot, Kirsten A., Steven M. Schneider, and Michael Xenos. "Online Campaigning in the 2002 U.W. Elections." Working Paper version 2. (Earlier version presented at the Internet Research 3.0 Conference, Maastricht, The Netherlands, October 2002). Retrieved December 24, 2005 from <*http://politicalweb.info/publications/2002Working Paper.pdf*>.

Foot, Kirsten, Barbara Warnick, and Steven M. Schneider. "Web-Based Memorializing after September 11: Toward a Conceptual Framework." *Journal of Computer-Mediated Communication* 11.1 (2005): article 4. Retrieved December 28, 2005 <*http://jcmc.indiana.edu/vol11/issue1/foot.html*>.

Fox, Susannah. "Digital Divisions." *Pew Internet and American Life Project*. October 5, 2005. Pew Charitable Trusts. Retrieved April 21, 2006 <*http://www.pewinternet.org.offcampus. lib.washington.edu/PPF/r/165/report_display.asp*>.

Fraser, Nancy. "Rethinking the Public Sphere: A Contribution to the Critique of Actually Existing Democracy." In *Habermas and the Public Sphere*. Ed Craig Calhoun. Cambridge: The MIT Press, 1992. 109–142.

Fürsich, Elfriede and Melinda B. Robins. "Africa.com: The Self Representation of Sub-Saharan Nations on the World Wide Web." *Critical Studies in Media Communication* 19 (2002): 190–211.

Gallagher, Leigh. "About Face." *Forbes* (March 19, 2001): 178. *ProQuest*. Retrieved April 12, 2006 <*http://www.proquest.com*>.

Gaonkar, Dilip Parameshwar. "The Forum: Publics and Counterpublics: Introduction." *Quarterly Journal of Speech* 88 (2002): 410–412.

Gibaldi, Joseph. *MLA Handbook for Writers of Research Papers*. 6th edn. New York: The Modern Language Association, 2003.

Gibson, Jason and Alex Kelly, "Become the Media." *Arena Magazine* (October 2000): 10. *Infotrac*. Retrieved November 23, 2005 <*http://infotrac.galegroup.com*>.

Golden, James L. and Edward P. J. Corbett, Eds. *The Rhetoric of Blair, Campbell, and Whately*. New York: Holt, Rinehart & Winston, 1968.

"Governor Bush Welcomes You to His Virtual Campaign Headquarters." *George W. Bush for President*. March 3, 2000. Bush for President. *Intenet Archive*. Retrieved March 27, 2006 <*http://web.archive.org/web/20000301032023/www.georgewbush.com/georgelaura/ index.html*>.

Gronbeck, Bruce E. and Danielle R. Wiese. "The Repersonalization of Presidential Campaigning." *American Behavioral Scientist* 49.4 (2005): 520–534.

Gurak, Laura J. *Persuasion and Privacy in Cyberspace*. New Haven, CT: Yale University Press, 1997.

Gurak, Laura J. and John Logie. "Internet Protests, from Text to Web." In *Cyberactivism: Online Activism in Theory and Practice*. Ed. Martha McCaughey and Michael D. Ayers. New York: Routledge, 2003. 25–46.

Haas, Gretchen. "Subject to the System: The Rhetorical Construction and Constitution of Internet Candidates and Citizens in the 2004 U. S. Presidential Campaign." Diss. University of Minnesota, 2006.

Habermas, Jürgen. *The Structural Transformation of the Public Sphere: An Inquiry into a Category of Bourgeois Society*. Trans. Thomas Burger with the assistance of Frederick Lawrence. Cambridge: The MIT Press, 1989. Published originally in German as *Strukturwandel der Offentlicheit* in 1962.

Hall, Stuart. "Encoding/Decoding." In *Media and Cultural Studies: Key Works*. Eds. Gigi Durham and Douglas M. Kellner. Oxford: Blackwell, 2001. 166–176.

Hamelink, Cees J. "Human Rights in the Global Billboard Society." In *Shaping the Network Society: The New Role of Civil Society in Cyberspace*. Ed. Douglas Schuler and Peter Day. Cambridge: The MIT Press, 2004. 67–81.

Hayles, N. Katherine. "Deeper into the Machine: Learning to Speak Digital." *Computers and Composition* 19 (2002): 371–386. Retrieved October 27, 2005 from Elsevier <*http://www.sciencedirect.com*>.

Hayles, N. Katherine. *Writing Machines*. Cambridge: The MIT Press, 2002.

Henderson, Diedtra. "Vioxx Jury Tells Merck to Pay; One of Two Who Had Heart Attacks Given $4.5M, but the Other Awarded No Damages." *The Boston Globe* (April 6, 2006): D1, 3rd edn. *Lexis Nexis*. University of Washington. Retrieved April 8 <*http://www. lexisnexis.com*>.

Herman, David. "Toward a Transmedial Narratology." In *Narrative across Media: The Languages of Storytelling*. Ed. Marie-Laure Ryan. Lincoln: University of Nebraska Press, 2004. 47–61.

Herring, Susan C. "Slouching toward the Ordinary: Current Trends in Computer-Mediated Communication." *New Media & Society* 6.1 (2004): 26–36. Retrieved June 13, 2004. Sage. <*http://nms.sagepub.com*>.

Hickman, Paul Forrest. "Presidential Campaign 2004 Archiving." Email to Barbara Warnick. December 6, 2005.

Hitchon, Jacqueline C. and Jerzy O. Jura. "Allegorically Speaking: Intertextuality of the Postmodern Culture and Its Impact on Print and Television Advertising." *Communication Studies* 48.2 (1997): 142–158.

Inch, Edward S., Barbara Warnick, and Danielle Endres. *Critical Thinking and Commuinication: The Use of Reason in Argument*. 5th edn. Boston: Pearson, 2006.

"Indymedia's Frequently Asked Questions (FAQ)." Retrieved April 30, 2006 <*http://docs.indymedia.org/view/Global/FrequentlyAskedQuestionEn#hits*>.

Institute for Politics, Democracy, and the Internet (IDPI). *Online Campaigning 2002*. Washington, DC: IDPI, 2002. Retrieved December 30, 2002 <*http://democracyonline. org/primer2002.html*>.

Irwin, William. "Against Intertextuality." *Philosophy and Literature* 28.2 (2004): 227–242. *Project Muse*. Johns Hopkins University Press. University of Washington. Retrieved March 27, 2006 <*http://muse.jhu.edu*>.

Iser, Wolfgang. *The Act of Reading: A Theory of Aesthetic Response*. Baltimore: The Johns Hopkins University Press, 1978.

Jamieson, Kathleen Hall. *Dirty Politics: Deception, Distraction, and Democracy*. New York: Oxford University Press, 1992.

"Judge Blocks Notorious BIG Album." *BBC News*. March 19, 2006. Retrieved April 1, 2006 <*http://news.bbc.co.uk/2/hi/entertainment/4823028.stm*>.

Kahn, Richard and Douglas Kellner. "New Media and Internet Activism: From the 'Battle of Seattle' to Blogging." *New Media and Society* 6 (2004): 87–95. Sage. Retrieved June 13, 2005 <*http://nms.sagepub.com*>.

Kampf, Constance E. "Kumeyaay Online: Dimensions of Rhetoric and Culture in the Kumeyaay Web Presence." Diss. University of Minnesota, 2005. *ProQuest*. <*http:// proquest.umi.com*> Retrieved April 21, 2006.

Kaplan, Nancy. "Literacy beyond Books: Reading When All the World's a Web." In *The World Wide Web and Contemporary Cultural Theory*. Ed. Andrew Herman and Thomas Swiss. New York: Routledge, 2000. 207–234.

Kawamoto, Dawn. "Microsoft, Gateway Reach Antitrust Settlement." *CNET News.com* April 11, 2005. Retrieved March 27, 2006 <*http://news.com.com/ Microsoft,+Gateway +reach+antitrust+settlement/2100-1014_3-5662409.html*>.

Kay, Russell. "Flash." *Computer World* 40.3 (2006): 32.

Kellner, Douglas. "Theorizing Globalization." *Sociological Theory* 20 (2002): 286–305.

Kennedy, George A., Ed. and Trans. *Aristotle, on Rhetoric*. New York: Oxford University Press, 1991.

Kennedy, George. *The Art of Persuasion in Greece*. Princeton, NJ: Princeton University Press, 1963.

Kever, Jeannie. "Bush Twins Host Online Chat; Sisters Answer 14 Questions in the Hourlong Session." *The Houston Chronicle*, 3 star edn: B7. Retrieved July 24, 2004 from *Lexis Nexis* <*http://www.lexisnexis.com/*>.

Kidd, Dorothy. "Indymedia.org: A New Communications Commons." *Cyberactivism: Online Activism in Theory and Practice*. New York: Routledge, 2003.

Killoran, John B. "@ Home among the .coms: Virtual Rhetoric in the Agora of the Web." In *Alternative Rhetorics: Challenges of the Rhetorical Tradition*. Ed. Laura Gray-Rosendale and Sibylle Gruber. Albany: State University of New York Press, 2001. 127–144.

Kiousis, Spiro. "Interactivity: A Concept Explication." *New Media & Society* 4 (2002): 355–383. Sage. Retrieved December 22, 2005 <*http://www.sagepub.com/*>.

Kranich, Nancy, "Libraries: The Information Commons of Civil Society." In *Shaping the Network Society: The New Role of Civil Society in Cyberspace*. Ed. Douglas Schuler and Peter Day. Cambridge: The MIT Press, 2004. 279–299.

Kress, Gunther. "New Forms of Text, Knowledge, and Learning." *Computers and Composition* 22 (2005): 5–22. Elsevier. Retrieved July 22, 2005 <*http://www.sciencedirect.com*>.

Kristeva, Julia. *Desire in Language*. New York: Columbia University Press, 1980.

Kunst, H., D. Groot, P. M. Latthe, M. Latthe, and K. S. Kahn. "Accuracy of Information on Apparently Credible Websites: Survey of Five Common Health Topics." *British Medical Journal* 324 (March 9, 2002): 581–582. *Proquest*. Retrieved August 13, 2004 <*http://proquest.umi.com*>.

Lasn, Kalle. "Media Carta." *Peace Review* 11.1 (1999): 121–124. *Research Library*. Taylor & Francis. University of Washington. Retrieved April 10, 2006 <*http://www.proquest. com*>.

"This Land Is Your Land." *Wikipedia*. 2006. Retrieved April 1, 2006. <*http://en.wikipedia. org/wiki/This_Land_Is_Your_Land*>.

Lemi, Baruh. "Music of My Own? The Tranformation from Usage Rights to Usage Privileges in Digital Media." In *Digital Media: Transformations in Human Communication*. Ed. Paul Messaris and Lee Humphreys. New York: Peter Lang Publishing, 2006. 67–78.

Lemke, Jay L. "Travels in Hypermodality." *Visual Communication* 1.3 (2002): 299–325.

Lieberman, Trudy. "Bitter Pill." *Columbia Journalism Review* (2005): n.p. Retrieved April 8, 2006 <*http://www.cjr.org/issues/2005/4/lieberman.asp*>.

"Live Tour." 2006. Ableton. Retrieved April 7, 2006 <*http://www.lexisnexis.com/default.asp*>.

Lortie, Bret. "Web Watch." *Bulletin of Atomic Scientists* 56.5 (2000): 9. *Proquest*. Retrieved August 24, 2005 <*http://proquest.umi.com*>.

"The Lowest Moments in Advertising." *Adweek* (June 9, 2003): 38. *ProQuest*. University of Washington. Retrieved April 12, 2006 <*http://www.proquest.com*>.

Lydon, Christopher. "After New Hampshire." Online blog. January 29, 2004. Retrieved August 22, 2005 <*http://www.bopnews.com/archives/000231.html*>.

Lynch, Patrick J. and Sarah Horton. *Web Style Guide: Basic Design Principles for Creating Web Sites*. 2nd edn. New Haven, CT: Yale University Press, 2001.

Malone, Julia. "Election 2004: People Nationwide to Party for Bush Today." *The Atlanta Journal-Constitution* (April 29, 2004), home edn: 1B. Retrieved December 14, 2005 from *Lexis Nexis* <*http://www.lexisnexis.com/*>.

Malone, Julia . "Election Awash in Money Despite New Campaign Finance Law." *Cox News Service* (November 5, 2004). Retrieved August 26, 2006 from *Lexis Nexis* <*http://www.lexisnexis.com/*>.

Maney, Kevin. "This Net Was Made for You and Me and the Rest of the World." *USA Today* (July 28, 2004). Retrieved April 7, 2006 from *Lexis Nexis* <*http://www.lexisnexis.com*>.

Manovich, Lev. *The Language of New Media*. Cambridge: The MIT Press, 2001.

"Marlboro Man." *Wikipedia*. 2006. Retrieved April 1, 2006 <*http://en.wikipedia.org/wiki/Marlboro_Man*>.

McCaughey, Martha and Michael D. Ayers, Eds. *Cyberactivism: Online Activism in Theory and Practice*. New York: Routledge, 2003.

McChesney, Robert W. *Rich Media, Poor Democracy: Communication Politics in Dubious Times*. Urbana: University of Illinois Press, 1999.

McCorkle, Ben. "Harbingers of the Printed Page: Nineteenth-Century Theories of Delivery as Remediation." *Rhetoric Society Quarterly* 35 (2005): 25–49.

McCoy, Adrian. "Elect to Make a Web-Informed Choice." *Pittsburgh Post-Gazette* (January 11, 2004), five star edn: G-1. Retrieved December 14, 2005 from *Lexis Nexis* <*http://www.lexisnexis.com/*>.

McLuhan, Marshall and Bruce R. Powers. *The Global Village: Transformations in World Life and Media in the 21st Century*. New York: Oxford University Press, 1989.

McMillan, Sally J. "Exploring Models of Interactivity from Multiple Research Traditions: Users, Documents, and Systems." In *The Handbook of New Media*. Ed. L. Lievrouw and S. Livingston. Thousand Oaks, CA: Sage, 2002. 163–182.

Meyrowitz, Joshua. "Medium Theory." In *Communication Theory Today*. Ed. David Crowley and David Mitchell. Cambridge: Polity Press, 2004. 50–77.

Minerva: Election 2002 Web Archive Browse. November 1, 2005. Library of Congress. Retrieved April 30, 2006 <*http://lcweb4.loc.gov/elect2002/*>.

Morton, Judy and C. Kay Weaver. "The Epistemic Struggle for Credibility: Rethinking Media Relations." *Journal of Communication Management* 9 (2005): 246–255. *Proquest.* Retrieved August 24, 2005 <*http://proquest.umi.com*>.

Nagourney, Adam. "Politics Faces Sweeping Change via the Web." *The New York Times* (April 2, 2006), Late edn: 1-1. Retrieved April 2, 2006 from *Lexis Nexis.*

Natharius, David. "The More We Know, the More We See: The Role of Visuality in Media Literacy." *American Behavioral Scientist* 48.2 (2004): 238–247. *Expanded Academic ASAP.* Gale Group. University of Washington. Retrieved April 17, 2006 <*http://infotrac.galegroup.com/default*>.

National Endowment for the Arts. *Reading at Risk: A Survey of Literary Reading in America.* Research Division Report #46. Washington, DC: National Endowment for the Arts, 2004. Retrieved October 30, 2005 <*http://www.arts.gov/pub/ReadingAt Risk.pdf* >.

Neal, Terry M. and Lois Romano. "Bush's Style as Issue of Substance; Voters' View of Personality Traits Worries Some Supporters." *The Washington Post* (April 2, 2000), final edn: A01. Retrieved April 1, 2006 <*http://www.lexisnexis.com/default.asp*>.

Newhagen, John E. "Interactivity, Dynamic Symbol Processing, and the Emergence of Content in Human Communication." *The Information Society* 20 (2004): 395–400.

Newton, Julianne H. "Influences of Digital Images on the Concept of Photographic Truth." In *Digital Media: Transformations in Human Communication.* Ed. Paul Messaris and Lee Humphreys. New York: Peter Lang, 2006. 3–14.

Nolan, Hamilton. "Group Sells Sneaker to Tread on Nike Turf." *PRweek* (September 6, 2004) [New York], u.s edn: 5. *ProQuest.* University of Washington. Retrieved April 14, 2006 <*http://www.proquest.com*>.

O'Keefe, Daniel J. *Persuasion: Theory and Research.* 2nd edn. Thousand Oaks, CA: Sage, 2002.

Olson, Kathryn M. and Clark D. Olson. "Beyond Strategy: A Reader-Centered Analysis of Irony's Dual Persuasive Uses." *Quarterly Journal of Speech* 90 (2004): 24–52.

Ong, Walter J. *Orality and Literacy: The Technologizing of the Word.* London: Methuen, 1982.

Ong. Walter J. *Ramus: Method and the Decay of Dialogue; From the Art of Discourse to the Art of Reason.* Cambridge, MA: Harvard University Press, 1958.

Ono, Kent A. "Rev. of *Alternative Rhetorics: Challenges to the Rhetorical Tradition.*" Ed. Laura Gray-Rosendale and Sibylle Gruber. *Quarterly Journal of Speech* 88 (2002): 461–462.

Opel, Andy and Rich Templin. "Is Anybody Reading This? Indymedia and Internet Traffic Reports." *Transformations: Online Journal of Region, Culture, and Society* (2005). Retrieved April 30, 2006 <*http://transformations.cqu.edu.au/journal/issue_10/article_08.shtml*>.

Ott, Brian and Cameron Walter. "Intertextuality: Interpretive Practice and Textual Strategy." *Critical Studies in Media Communication* 17.4 (2000): 429–446.

Owens, Lynn and L. Kendall Palmer. "Making the News: Anarchist Counter-Public Relations on the World Wide Web." *Critical Studies in Media Communication* 20.4 (2003): 335–361.

Papacharissi, Zizi. "The Virtual Sphere: The Internet as a Public Sphere." *New Media & Society* 4 (2002): 9–27. Sage. June 13, 2005 <*http://nms.sagepub.com*>.

"Parody." *Cambridge Dictionaries Online.* 2006. University of Washington. Retrieved March 21, 2006 <*http://dictionary.cambridge.org/define.asp?key=57663&dict=CALD*>.

Perelman, Chaïm and Lucie Olbrechts-Tyteca. *The New Rhetoric: A Treatise on Argumentation.* Trans. John Wilkenson and Purcell Weaver. Notre Dame: University of Notre Dame Press, 1969.

Perlmutter, David. "Hypericons: Famous News Images in the Internet-Digital-Satellite Age." In *Digital Media: Transformations in Human Communication.* Ed. Paul Messaris and Lee Humphreys. New York: Peter Lang, 2006. 51–64.

Pickard, Victor W. "Assessing the Radical Democracy of Indymedia: Discursive, Technical, and Institutional Considerations." *Critical Studies in Media Communication* 23.1 (2006): 19–38.

Pollack, Andrew. "Drug Makers Nag Patients to Stay the Course." *The New York Times* (March 11, 2006), final edn: C1. *Lexis Nexis.* University of Washington. Retrieved April 8, 2006 <*http://www.lexisnexis.com/default.asp*>.

Potter, Trevor. "McCain-Feingold: A Good Start." *The Washington Post* (June 23, 2006), final ed.: A25. *Lexis Nexis.* University of Washington. June 25, 2006.

"Presidential Bloopers: Bush Spoke Better This Year, but His Talent for Malapropisms Didn't Desert Him." *Pittsburgh Post-Gazette* (December 28, 2003), two star edn: E-7. *Lexis Nexis.* Retrieved April 1, 2006 <*http://www.lexisnexis.com/default.asp*>.

Prosise, Theodore O., Jordan P. Mills, and Greg R. Miller. "Fields as Arenas of Practical Discursive Struggle: Argument Fields and Pierre Bourdieu's Theory of Social Practice." *Argumentation & Advocacy* 32.3 (1996): 111–128.

Puopolo, Sonia "Tita." "The Web and U.S. Senatorial Campaigns." *American Behavioral Scientist* 44 (2001): 2030–2047.

Rabinow, Paul. Ed. *The Foucault Reader.* New York: Pantheon, 1984.

Rafaeli, Sheizaf. "Interactivity: From New Media to Communication." In *Sage Annual Review of Communication Research.* Ed. R. P. Hawkins, J. M. Weimann, and S. Pingree. Beverly Hills, CA: Sage, 1988. 110–134. Retrieved March 5, 2004 <*http://sheizaf.rafaeli.net*>.

Rainie, Lee, Michael Cornfield, and John Horrigan. *The Internet and Campaign 2004.* Washington, DC: Pew Internet and American Life Project, 2005. Retrieved December 24, 2005 <*http://www.pewinternet.org/PPF/r/150/report_display.asp*>.

"Remarks by Governor Bush Lite." *DieTryin.com.* 2004. Retrieved July 7, 2000 <*http://www.bushlite.net/speech.html*>.

Rheingold, Howard. *The Virtual Community: Homesteading on the Electronic Frontier.* New York: HarperCollins, 1993.

Rieh, Soo Young. "Judgment of Information Quality and Cognitive Authority in the Web." *Journal of the American Society for Information Science and Technology* 53.2 (2002): 145–161.

Rumbo, Joseph D. "Consumer Resistance in a World of Advertising Clutter; the Case of *Adbusters.*" *Psychology & Marketing* 19.2 (2002): 127–148.

Ryan, Marie-Laure. "Introduction." In *Narrative across Media: The Languages of Storytelling.* Ed. Marie-Laure Ryan. Lincoln: University of Nebraska Press, 2004. 1–35.

Samuel, Alexandra. "Internet Plays Wild Card into U.S. Politics." *The Toronto Star* (October 18, 2004): DO1. Retrieved December 16, 2005 from *Lexis Nexis* <http://www.lexisnexis.com/>.

"Satire." *Oxford English Dictionary.* 2nd edn. 1989. University of Washington. Retrieved March 21, 2006 <http://dictionary.oed.com.offcampus.lib.washington.edu/cgi/entry/50213751?query_type=word&queryword=satire&first=1&max_to_show=10&sort_type=alpha&result_place=1&search_id=SOoD-7uPS8E-8048&hilite=5021>.

Schachter, Ken. " 'Online Political Satire Draws More Visitors than Candidates' Web.' " *Long Island Business News* (August 27, 2004). *Find Articles.* findarticles.com. Retrieved April 7, 2006 <http://findarticles.com/p/articles/mi_qn4189/is_20040827/ai_n10170650#continue>.

Schlein, Alan M. *Find It Online: The Complete Guide to Online Research.* Tempe, AZ: Facts on Demand Press, 2003.

Schneider, Steven, Kirston Foot, Alex Halavais, Adrienne Massanari, Elena Larson, Erica Siegl, and Megan Dogherty. *One Year Later: September 11 and the Internet.* Washington, DC: Pew Internet and American Life Project, September 5, 2002. Retrieved December 28, 2005 <http://207.21.232.103/PPF/r/69/report_display.asp>.

Scholes, Robert. *Protocols of Reading.* New Haven, CT: Yale University Press, 1989.

Schuler, Douglas and Peter Day, Eds. *Shaping the Network Society: The New Role of Civil Society in Cyberspace.* Cambridge: The MIT Press, 2004.

Seelye, Katherine Q. "At Newspapers, Some Clipping." *The New York Times* (October 10, 2005): C1, C5.

Selfe, Cynthia L. *Technology and Literacy in the Twenty-First Century: The Importance of Paying Attention.* Carbondale: Southern Illinois University Press, 1999.

Skelton, Chad. "Adbusters Takes Run at Nike with Line of Logo-Free Shoes." *The Vancouver Sun* (September 7, 2004), final edn: A3. *ProQuest.* University of Washington. Retrieved April 14, 2006 <http://www.proquest.com>.

"SnagIT Screen Capture and Sharing." 2006. TechSmith. Retrieved April 7, 2006 <http://techsmith.com/snagit.asp>.

Soffer, Oren. "The Textual Pendulum." *Communication Theory* 15.3 (2005): 266–291.

"Sorry, Dick!." *TooStupidToBePresident.com.* n.d. Retrieved March 27, 2006 <http://www.toostupidtobepresident.com/shockwave/sorrycheney.htm>.

Souheil, Marine and Jean-Marie Blanchard. *Bridging the Digital Divide: An Opportunity for Growth for the 21st Century.* Paris: Compagnie Financière Alcatel, 2005. Retrieved October 30, 2005 <http://www.alcatel.com/doctypes/articlepaperlibrary/html/ATR2004Q3/ATR2004Q3A02_EN.jhtml#top>.

Souley, Boubacar and Robert H. Wicks. "Tracking the 2004 Presidential Campaign Web Sites." *American Behavioral Scientist* 49.4 (2005): 535–547.

Stein, Sarah R. "The '1984' Macintosh Ad: Cinematic Icons and Constitutive Rhetoric in the Launch of a New Machine." *Quarterly Journal of Speech* 88.2 (2002): 169–192.

Strauss, Robert. " 'Brothers' Cartoon Parody Becomes Most-Watched Campaign Item." *Philadelphia Daily News* (October 27, 2004): n.p. *Lexis Nexis*. University of Washington. Retrieved April 7, 2006 <*http://www.lexisnexis.com/ default.asp*>.

Stromer-Galley, Jennifer. "On-Line Interaction and Why Candidates Avoid It." *Journal of Communication* 50 (2000): 111–132.

Stromer-Galley, Jennifer and Kirsten A. Foot. "Citizen Perceptions of Online Interactivity and Implications for Political Campaign Communication." *Journal of Computer-Mediated Communication* 8 (2002). Retrieved February 6, 2004 <*http://jcmc.indiana.edu/vol8/issue1/stromerandfoot.html*>.

"Sweatshops: Frequently Asked Questions." *Global Exchange*. 2005. Global Exchange. Retrieved April 14, 2006 <*http://www.globalexchange.org/campaigns/sweatshops/nike/faq.htm*>.

Toulmin, Stephen Edelston. *The Uses of Argument*. Cambridge: Cambridge University Press, 1969.

Toulmin, Stephen, Richard Rieke, and Allan Janik. *An Introduction to Reasoning*. 2nd edn. New York: Macmillan, 1984.

Tyner, Kathleen. *Literacy in a Digital World: Teaching and Learning in the Age of Information*. Mahwah, NJ: Lawrence Erlbaum Associates, 1998.

United Colors of Benetton. 2006. Benetton Group. Retrieved April 12, 2006 <*http://www.benetton.com/html/index.shtml*>.

Walji, M. S. Sagaram, D. Sagaram, F. Merc-Bernstam, C. Johnson, N. O. Mirza, and E. V. Bernstam. "Efficacy of Quality Criteria to Identify Potentially Harmful Information: A Cross-Sectional Survey of Complementary and Alternative Medicine Web Sites." *Journal of Medical Internet Research*, 6.2 (2004): Art. No. e21. Retrieved January 22, 2005 from the Web of Science database.

Warner, Michael. "Publics and Counterpublics." *Public Culture* 14 (2002): 49–90. *Project Muse* (August 19, 2005) <*http://muse.jhu.edu*>.

Warnick, Barbara. "Appearance or Reality? Political Parody on the Web in Campaign '96." *Critical Studies in Mass Communication* 15 (1998): 306–324.

Warnick, Barbara. *Critical Literacy in a Digital Era: Technology, Rhetoric, and the Public Interest*. Mahwah, NJ: Lawrence Erlbaum Associates, 2002.

Warnick, Barbara. *The Sixth Canon: Belletristic Rhetorical Theory and Its French Antecedents*. Columbia: University of South Carolina Press, 1993.

Warnick, Barbara, Michael Xenos, Danielle Endres, and John Gastil. "Effects of Campaign-to-User and Text-Based Interactivity in Political Candidate Campaign Web Sites." *Journal of Computer-Mediated Communication* 10.3 (2005): Article 5. Retrieved December 24, 2005 <*http://jcmc.indiana.edu/vol10/issue3/warnick.html*>.

Wathen, C. Nadine and Jacquelyn Burkell. "Believe It or Not: Factors Influencing Credibility on the Web." *Journal of the American Society for Information Science and Technology* 53.2 (2002): 134–144.

"What is MoveOn?" *MoveOn.org Democracy in Action*. n.d. Moveon.org. Retrieved June 24, 2006 <*http://www.moveon.org/about.html*>.

"What We Do: The Issues We Work on Worldwide." *Greenpeace International*. n.d. Retrieved August 24, 2005 <*http://www.greenpeace.org/international/campaigns*>.

"Who Owns the Media?" *Freepress.net*, n.d. Retrieved August 16, 2005. <*http://www.freepress.net/content.ownership*>.

Wiese, Danielle R. and Bruce E. Gronbeck. "Campaign 2004 Developments in Cyberpolitics." In *The 2004 Presidential Campaign: A Communication Perspective*. Ed. Robert E. Denton, Jr. Lanham, MD: Rowman & Littlefield, 2005. 217–239.

Williams, Kate and Abdul Alkalimat. "A Census of Public Computing in Toledo, Ohio." In *Shaping the Network Society: The New Role of Civil Society in Cyberspace*. Ed. Douglas Schuler and Peter Day. Cambridge: The MIT Press, 2004. 85–110.

Williams, Paul Andrew. "The Main Frame: Assessing the Role of the Internet in the 2004 Presidential Contest." In *The 2004 Presidential Campaign: A Communication Perspective*. Ed. Robert E. Denton, Jr. Lanham, MD: Rowman & Littlefield, 2005. 241–254.

Wolf, Gary. "How the Internet Invented Howard Dean." *Wired Online* 12.1 (August 22, 2005). <*http://www.wired.com*>.

Zun, L. S., D. N. Blume, J. Lester, G. Simpson, and L. Downey. "Accuracy of Emergency Medical Information on the Web." *American Journal of Emergency Medicine* 22.2 (2004): 94–97. Retrieved January 22, 2005 from the Web of Science database.

Index